'X' STANDS FOR UNKNOWN

ISAAC ASIMOV

 A DISCUS BOOK/PUBLISHED BY AVON BOOKS

The following essays in this volume are reprinted from *The Magazine of Fantasy and Science Fiction*, having appeared in the indicated issues:

The Long Ellipse (January 1982)
The Circle of the Earth (February 1982)
Whatzisname's Orbit (March 1982)
Change of Time and State (April 1982)
Read Out Your Good Book in Verse (May 1982)
Four Hundred Octaves (June 1982)
The Three Who Died Too Soon (July 1982)
X Stands for Unknown (August 1982)
Big Brother (September 1982)
Bread and Stone (October 1982)
A Difference of an "E" (November 1982)
Silicon Life After All (December 1982)
To Ungild Refined Gold (January 1983)
Ready and Waiting (February 1983)
The Armies of the Night (March 1983)
Dead Center (April 1983)
Out in the Boondocks (May 1983)

AVON BOOKS
A division of
The Hearst Corporation
1790 Broadway
New York, New York 10019

The Doubleday edition contains the following Library of Congress Cataloging in Publication Data:

Asimov, Isaac, 1920–
X stands for unknown.
Essays selected from the magazine of fantasy and science fiction.
1. Science—Addresses, essays, lectures.   2. Mathematics—Addresses, essays, lectures. I. Magazine of fantasy and science fiction.   II. Title.
Q171.A73   1984   500   83-45019

First Discus Printing, March, 1985

DISCUS TRADEMARK REG. U.S. PAT. OFF. AND IN
OTHER COUNTRIES, MARCA REGISTRADA, HECHO EN U.S.A.

Printed in the U.S.A.

OPB   10   9   8   7   6   5   4   3   2   1

Contents

Introduction

When I was in my early teens and still in high school, the local pharmacist (not, as I look back on it, a very intelligent man) undertook to prove to me the presence of divine power by a simple demonstration.

He said to me, "Scientists can't even synthesize sucrose, something almost any plant can do."

I was amazed. "So what?" I said. "There are a million things scientists can't do and don't know. What's that got to do with it?"

The pharmacist, however, overbore me and insisted that the failure to synthesize sucrose (which I imagine chemists *can* synthesize, by the way) proved the existence of a supernatural entity. I was sufficiently young to be unsure of myself and to be abashed in the presence of an adult, so I forbore to continue the argument—but I was in no way convinced by him, nor was my own view of the matter shaken in the slightest.

It is a common mistake. There seems to be a vague notion that something omniscient and omnipotent *must* exist. If it can be shown that scientists are not all-knowing and all-powerful, then that must be the proof that something else that *is* omniscient and omnipotent *does* exist. In other words: Since scientists can't synthesize sucrose, God exists.

Well, God may exist; I won't argue the point here—but this type of argument does not suffice to prove the matter. In fact, such an argument can be advanced only by people who miss the whole point of science.

Science is *not* a collection of results, abilities, or even explanations. Those are products of science, but are not science itself—any more than a table is carpentry, or standing at the finish line is racing.

The results, abilities, and explanations produced by science are tentative, and are possibly wrong in whole or in part. They are almost certainly incomplete. None of that necessarily implies any flaw or insufficiency in science itself.

Science is a *process;* it is a way of thinking, a manner of approaching and of possibly resolving problems, a route by which one can produce order and sense out of disorganized and chaotic observations. Through it we achieve useful conclusions and results that are compelling and upon which there is a tendency to agree. These scientific conclusions are commonly looked upon as representing a reasonable approach to ''truth''—subject to later emendation.

Science does not promise absolute truth, nor does it consider that such a thing necessarily exists. Science does not even promise that everything in the Universe is amenable to the scientific process.

Science deals only with those portions and conditions of the Universe that can be reasonably observed and for which the tools it uses are adequate. The tools (including such immaterial ones as mathematics and logic) may be improved with time, but there is no guarantee that they can be improved indefinitely, so as to overcome all limits.

Furthermore, even when dealing with matters amenable to observation and analysis, science cannot guarantee that a reasonable solution will come about in any set time. One may be long delayed for lack of a key observation, or of an appropriate flash of insight.

The process of science, therefore, involves a slow forward movement through the reachable portions of the Universe—a gradual unfolding of parts of the mystery.

The process may never be finished. There may never be a time when all mysteries are resolved, when nothing remains to be done within the field that the scientific process is competent to deal with. Consequently, at any fixed moment—say, now—there are unsolved problems, and this proves nothing with respect to God, one way or the other.

Nor should this eternal perpetuation of mystery be a source of disappointment, it seems to me. It should, rather, be a source of over-whelming relief. If all questions were answered, all riddles solved, every fold unfolded, every wrinkle in the fabric of the Universe smoothed—the greatest and noblest game in the Universe would be ended, and there would be nothing left for the mind to do but to console itself with trivia.

Unbearable.

If we assume the existence of an omniscient and omnipotent being, one that knows and can do absolutely everything, then to my own very limited self, it would seem that existence for it would be unbearable. Nothing to wonder about? Nothing to ponder over? Nothing to discover? Eternity in such a heaven would surely be indistinguishable from hell.

A few years ago I wrote a story about an omniscient, omnipotent being (and therefore an eternal one) who had created a Universe so designed as to give rise to uncounted forms of intelligent life. He then collected huge numbers of such life forms and put them all to the task of making new discoveries, in the perhaps hopeless hope that one of them might find the being to be not *quite* omniscient, and might therefore come up with a method (unknown to the being) of lifting the insupportable burden of immortality from his shoulders.

Well, then, because of my belief that the true delight is in the finding out, rather than in the knowing, my tendency in writing my science essays is not to describe knowledge flatly, as though it were delivered from some source of all knowledge, but, instead, when I can, to describe that manner in which what is known became known; how it was found out, step by step.

And so I found a name for this particular collection.

In the course of the last seventeen months, I wrote a four-part series of essays on the electromagnetic spectrum. (As is usual in such cases, I had fondly supposed I would be able to dispose of the matter in a single essay, but these essays write themselves and I have little to do with it.)

The fourth of these essays I entitled "X Stands for Unknown," for reasons that will be perfectly plain when you read it. However, meditating on the uses of the unknown, the delights of tackling it, and the relief of finding that it will not go away no matter how successful we are in the tackling, I decided to shift the title to the book as a whole.

May X always be with us to afford us pleasure.

PHYSICS

I

Read Out Your Good Book in Verse

The first mnemonic sentence I ever learned, when I was quite a small boy, was "Read Out Your Good Book in Verse."

If you take the initial letters of these words—ROYGBIV—you get the seven colors, in order, that Isaac Newton (1642–1727) reported in the optical spectrum: Red, Orange, Yellow, Green, Blue, Indigo, and Violet.

I was delighted beyond words at this discovery; not so much at the spectrum, which seemed perfectly straightforward to me, but by the existence of mnemonic sentences. It had never occurred to me that such a thing was possible, and for a while I thought I had the key to all knowledge.

Just invent enough mnemonic sentences, I thought, and you won't ever have to memorize anything again.

Unfortunately, as I was later to discover in almost all the great ideas I was ever to have, there was a fatal catch. You had to memorize the sentences and they were just as hard to remember as the raw data— even harder. For instance, to this day, I don't really have "Read Out Your Good Book in Verse" memorized. The way I get it is to think of the colors of Newton's spectrum in order (something I have no trouble in doing) and then work out the mnemonic sentence from the initial letters of the colors. That's what I had to do when I started this essay.

I had a different kind of trouble with the mnemonic, too. It wasn't accurate. I occasionally came across books which showed me colored pictures of the optical spectrum and I had no trouble seeing red at one end, and then following it down through orange, yellow, green, and blue.

Beyond blue I had a problem. What I saw at the other end of the spectrum was a color I called "purple." (Actually, I called it "poiple" as all decent Brooklyn kids did, but I knew it was *spelled* "purple" for some arcane reason.)

That wasn't fatal. I was willing to accept "violet" as a la-di-da fancy-shmancy synonym for "purple," on a par with "tomahto" and "eyether." And I could always modify the mnemonic into "Read Out Your Good Book in Prose."

What bothered me much more, however, was that I saw no color between blue and violet. My eye could make out nothing that I could identify as "indigo." Nor could anyone I consulted see this mysterious color. The best I could get out of anyone was that indigo was blue-purple. In that case, though, why wasn't blue-green a separate color?

Finally, I said, "The heck with it!" and left it out. I changed the mnemonic sentence to "Read Out Your Good Book, Victor" (or "Read Out Your Good Book, Peter"). What's more, I can find no modern physics texts that list indigo among the colors of the spectrum. They list six colors and no more.

Nevertheless, such is the force of tradition that, some twenty years ago, when I wrote an essay on the spectrum for a Minneapolis newspaper, and made no reference to indigo, I received several letters denouncing me quite vigorously for having omitted a color.

Nevertheless, I will continue to do so in this essay.

I described how Newton first obtained the light spectrum in 1666 in my essay "The Bridge of the Gods" (see *The Planet That Wasn't*, Doubleday, 1976). The existence of the spectrum did not in itself, however, indicate the nature of light. Newton himself thought light consisted of a spray of ultratiny particles, traveling in straight lines. He reasoned this was so from the fact that light cast sharp shadows. If light consisted of waves, as a competing suggestion maintained, then light would be expected to curve about the edge of an obstacle and cast a fuzzy shadow or even no shadow at all. Water waves curved about obstacles, after all, and sound, which was strongly suspected of consisting of waves, did the same.

Newton's contemporary, the Dutch scientist Christian Huygens (1629–95), was the chief supporter of the notion of light waves and he maintained that the shorter the wave, the smaller the tendency to curve about obstacles. Sharp-edged shadows were, in that case, not inconsistent with the wave notion provided the waves were short enough.

As a matter of fact, in a book published posthumously in 1665, an Italian physicist, Francesco Maria Grimaldi (1618–63), reported experiments in which he found that shadows were not perfectly sharp-edged and that light *did* bend, very slightly, about obstacles.

Newton knew of this experiment and tried to explain it in terms of particle theory. His successors, however, convinced that Newton could do no wrong, and that if he said "particles" it was particles, simply ignored Grimaldi.

Finally, in 1803, the English scientist Thomas Young (1773–1829) swung the weight of opinion to the side of waves. He passed light through two narrow orifices in such a way that the beams, as they emerged, overlapped on a screen. The overlapping did not simply increase the light upon the screen. Instead, it produced alternating bands of light and darkness.

If light consisted of particles, there was no way of explaining the appearance of bands of darkness. If it consisted of waves, then it was easy to see that there were conditions under which some of the waves might be moving upward as others were moving downward, and the two displacements would cancel each other, leaving nothing. In this way the two patches of light "interfered" with each other, and the bands of light and darkness were called "interference fringes."

This phenomenon is well known in the case of sound, and produces something called "beats." Interference fringes are the optical analogs of sonic beats.

From the width of the interference fringes, Young was able to make the first estimate of the length of the light waves and decided they were in the range of 1/50,000th of an inch, which is correct. He determined the wavelength of each of the colors and showed, with reasonable accuracy, the manner in which the wavelengths decreased from red to violet.

Of course, while wavelengths are a physical reality, colors are not. Anyone, given the proper instruments and training, can determine the wavelength of a particular variety of light wave. Determining its color, however, depends on the individual response of the pigments in the retina, and the interpretation that the brain makes of that response.

Different retinas might not be absolutely alike in their response to a particular wavelength. Some eyes, which are defective in certain retinal pigments, are partially or entirely color-blind. And even if two people detect color with equal sensitivity, how can anyone compare the mental interpretation? You cannot describe what you see when you see red, except by pointing to something that gives you the impression of red. Someone else will agree that it gives *him* the impression he, too, has been taught to call red, but how can you possibly tell whether your impression and his are identical?

Two people may forever agree on what to *call* the color of every object and yet forever *see* different things altogether. And no one can possibly explain color to someone who has been blind from birth so that the possibility of pointing to something and saying "This is red" is nonexistent.

What's more, as one goes along the spectrum, seeing only one wavelength's worth at a time, so to speak, there is no sharp change from red to orange, or from orange to yellow. There is a very slow and gradual shift, and there is absolutely no way of saying that "at this point the color is no longer red, but is orange."

If you were to move along the wavelength range and ask each of a great many people to indicate where the color has definitely ceased being orange and become yellow, you are sure to get a scattering. Different people will indicate slightly different wavelengths. Textbooks that therefore give limits, and say yellow stretches from one particular wavelength to another, are being misleading.

I think it is better to give a wavelength that is in the middle of the range of each color, a wavelength on which people with normal retinas will agree to call red, or green, or whatever.

Light wavelengths have traditionally been given in "Angstrom units," named in 1905 for the Swedish physicist Anders Jonas Ångström (1814–74), who first used them in 1868. An Angstrom unit is one ten-billionth of a meter, or 1×10^{-10} meters.

Nowadays, however, it is considered bad form to use Angstrom units because they disrupt the regularity of the metric system. It is considered preferable now to use different prefixes for every three orders of magnitude, with "nano" the accepted prefix for a billionth (10^{-9}) of a unit.

In other words, a "nanometer" is 10^{-9} meters, so that 1 nanometer equals 10 Angstrom units. If a particular light wave has a wavelength of 5,000 Angstrom units, it also has a wavelength of 500 nanometers, and it is the latter that should be used.

Here, then, are the midrange wavelengths of the six colors of the spectrum:

color	wavelength (nanometers)
red	700
orange	610
yellow	575
green	525
blue	470
violet	415

How long can a wavelength be and still produce a color that is detectably red to the eye; how short and still produce a color that is detectably violet? This varies from eye to eye, but the longest red wavelength, as seen by normal eyes before it fades to blackness, is usually given as 760 nanometers, and the shortest violet wavelength as 380 nanometers.

Although Thomas Young himself invented the term "energy" in 1807, it was not till the middle of the nineteenth century that the conservation of energy was understood, and not till the beginning of the twentieth century that it was clear that the energy content of light increased as the wavelength decreased. Of the colors of the spectrum, in other words, red is the least energetic and violet the most.

It isn't immediately obvious (at least to me) why short-wave light is more energetic than long-wave light, but the situation improves if we look at the matter in another way.

In one second, a beam of light will travel 299,792,500 meters or, roughly, 3×10^8 meters. If the light that is traveling happens to have a wavelength of 700 nanometers (7×10^{-7} meters), then the number of individual waves that will fit into that one second's length of light is 3×10^8 divided by 7×10^{-7}, or about 4.3×10^{14}.

That is the "frequency" of the light, and what it means is that, in one second, light of wavelength 700 nanometers will vibrate 430 trillion times.

We can work out the frequency for the midrange of each color:

color	frequency (trillion)
red	430
orange	490
yellow	520
green	570
blue	640
violet	720

If we consider frequencies, it seems to me that the greater energy of the short-wave light becomes more apparent. The short waves vibrate more quickly. If you are shaking something, it will clearly take more energy to shake it quickly than to shake it slowly, and so the shaking object will contain more energy if it is vibrating quickly. So it is that the basic discovery of quantum theory is that there is a unit of energy of radiation ("quantum") that is proportional in size to the frequency of that radiation.

The longest red wavelength, and hence the least energetic bit of visible light, has a frequency of just about 4.0×10^{14}, or 400 trillion. The shortest violet wavelength, and hence the most energetic portion of visible light, has a frequency of just about 8.0×10^{14}, or 800 trillion.

As you see, the farthest visible stretch of violet light has just half the wavelength, and therefore twice the frequency and twice the energy, of the farthest visible stretch of red light.

When we are dealing with sound, we have notes going up the scale:

do, re, mi, fa, sol, la, ti, do. If we wish, we can repeat that several times in either direction. In going from any "do" to the next higher "do," we exactly double the frequency of the sound waves. And if we start with "do" as the first note, and continue to count notes as we go up the scale, then the eighth note is "do" again and we have doubled the frequency. For that reason, we call the stretch from "do" to "do" an "octave," from the Latin word for "eighth."

That notion is extended, so that any stretch of wave motion of any kind, from a particular frequency to double that frequency, is called an octave. Thus, the stretch of light waves from the extreme red to the extreme violet, with a range of frequency from 400 trillion to 800 trillion, is called an octave, even though light doesn't consist of notes, and certainly not eight of them. (If you want to draw an analogy between colors and notes—a very poor analogy—just remember that there are only six colors. Even if you resurrect indigo, you have only seven.)

Sound waves vary in pitch as the wavelength changes. The longer the wavelength (and the lower the frequency), the deeper the sound. The shorter the wavelength (and the higher the frequency) the shriller the sound. The deepest note the normal ear can hear is perhaps 30 vibrations per second. The highest note the ear can hear varies with age, for the upper limit recedes as one grows older. Children can hear sounds with a frequency of up to 22,000 per second.

If we begin with 30 and double it over and over, we find that after 9 doublings we reach a frequency of 15,360 per second. Another doubling will carry it past the shrillest sound a child can hear. Consequently, we can say that the human ear can hear sounds over a stretch of a little more than 9 octaves. (The 88 notes on the standard piano keyboard have a range of a little more than 7 octaves.)

In contrast, our eyes see light over a range of exactly one octave. This may make it sound as though vision is very limited compared to hearing, but light waves are much shorter and more energetic than sound waves and can carry correspondingly more information. The typical visible light frequency is about 500 billion times as high as the typical audible sound frequency, so without meaning to downgrade the importance of hearing, there can be no question that our primary method of obtaining information concerning our surroundings is through vision.

The next question we might ask is this: Is the one octave of light all there is, or merely all we see?

Through most of history such a question would have sounded silly. A person would take it for granted that light, by definition, is something you see. If you can't see any light, it is because no light is there. The thought of invisible light would seem as much a contradiction in terms as "a square triangle."

The first indication that "invisible light" was not a contradiction in terms came in 1800.

In that year, the German-British astronomer William Herschel (1738–1822), who was famous as the discoverer of Uranus two decades before (see "The Comet That Wasn't," in *Quasar, Quasar, Burning Bright,* Doubleday, 1978), was experimenting with the spectrum.

It was common knowledge that when sunlight fell upon you, you felt a sensation of warmth. The general feeling was that the sun radiated both light *and* heat, and that the two were separate.

Herschel was wondering whether the heat radiation was spread out in a spectrum as light was, and he thought he might draw some conclusion on the matter if he placed the bulb of a thermometer at different parts of the spectrum. Since yellow, in the middle of the spectrum, seems the brightest portion, he expected temperature to rise higher as one progressed from either end of the spectrum toward the middle.

That did not happen. What he noticed, instead, was that the temperature rose steadily as one progressed away from the violet and reached its maximum in the far red. Astonished, Herschel wondered what would happen if he placed the thermometer bulb *beyond* the red. He tried it and found, to his even greater astonishment, that the temperature was higher there than anywhere in the visible spectrum.

This was three years before Young's demonstration of the existence of light waves and, for a time, it seemed as though there were indeed light rays and heat rays that were refracted differently, and partially separated, by a prism.

For a while, Herschel talked of "colorific rays," those that produced color, and "calorific rays," those that produced "calor," which is Latin for "heat." This had the virtue of sounding cute, but not only was it a gauche mixture of English and Latin, but it simply begged for endless misunderstandings due to misreading and typographical errors. Fortunately, it didn't catch on.

Once Young's demonstration of light waves was accepted, it could be maintained that what existed beyond the red end of the spectrum were light waves that were longer than those of red, and of lower frequency. Such waves would be too long to affect the retina of the eye and were, therefore, invisible, but except for that, they might be expected to have all the physical properties of the waves making up the visible portion of the spectrum.

Eventually, such radiation was termed "infrared" radiation, the prefix "infra" coming from the Latin meaning "below." The term is apt, since the frequency of infrared light is below that of visible light.

This means that infrared light also has less energy than visible light, and it may seem strange, in that case, that the thermometer registered a higher figure in the infrared than in the visible portion of the spectrum.

The answer is that the energy content of light is not all that must be taken into account.

We now know that the heating effect of the Sun's radiation does not depend on a separate set of heat rays. Instead, it is the light itself that is absorbed by opaque objects (at least in part) and the energy of this absorbed light is converted into the random energy of atomic and molecular vibrations—which we sense as heat. The amount of heat we obtain depends not only on the energy content of the light, but on how large a fraction of the light we absorb, rather than reflect.

The longer the wavelength (at least in the visible part of the spectrum) the more penetrating the light and the more readily it is absorbed, rather than reflected. Hence, even though red light is less energetic than yellow light, the greater efficiency of red-light absorption is such that it overbalances the other effect (at least where Herschel's thermometer was concerned). It is for this reason that the red region of the spectrum raised Herschel's thermometer to a higher temperature than other portions of the spectrum did and why infrared raised it higher still.

All this makes perfectly good sense in hindsight, but even after Young's demonstration of light waves was accepted, the wave nature of infrared could not simply be taken for granted. It would be necessary to *demonstrate* that wave nature, and that was hard to do. Experiments that were perfectly plain, where visible light was involved, because you could *see* what was happening—you could see the interference fringes, for instance—would not work with "invisible light."

You might imagine, of course, that you could use a thermometer instead. If there were interference fringes of infrared radiation, you might not see them, but if you ran a thermometer bulb along the screen on which the radiation existed, you would find regions in which the temperature did not rise and regions in which it did, and these regions would alternate—and that would be it.

Unfortunately, ordinary thermometers were simply not delicate enough for the job. It took them a long time to absorb enough heat to reach an equilibrium temperature, and the bulb was too thick to fit inside the interference fringes. For about half a century, therefore, after the discovery of infrared radiation, there wasn't much to be done with it, for the lack of proper instrumentation.

But then, in 1830, an Italian physicist, Leopoldo Nobili (1784–1835), invented the "thermopile." This consisted of wires of different metals joined at both ends. If one end is placed in cold water and the other is heated, a small electric current is set up in the wire. The current increases with the temperature difference between the two ends.

The current is easily measured, and a thermopile measures temperature far more quickly and sensitively than an ordinary thermometer does. What's more, the business end of a thermopile is considerably smaller than the bulb of an ordinary thermometer. For these reasons, a thermopile can measure the temperature of a small region and follow the ups and downs of interference fringes, for instance, where an ordinary thermometer couldn't.

Working along with Nobili was another Italian physicist, Macedonio Melloni (1798–1854). He found that rock salt was particularly transparent to infrared radiation. He therefore manufactured lenses and prisms out of rock salt, and used them to study infrared.

Between his rock-salt equipment and a thermopile, Melloni was able to show that infrared radiation had all the physical properties of ordinary light. It could be reflected, refracted, polarized, and it could produce interference fringes from which its wavelength could be determined. In 1850, Melloni published a book summarizing his work, and from that time on it was clear that "invisible light" was not a contradiction in terms; that the light spectrum extended well beyond the one-octave visible range.

By 1880 an American astronomer, Samuel Pierpont Langley (1834–1906), went even further. He made use of refraction gratings rather than prisms, and these spread out the infrared radiation into a wider and more efficient spectrum. He also devised a temperature indicator called a "bolometer," which consisted, in essence, of a fine platinum wire blackened to increase the efficiency with which it absorbed heat. Even tiny increases in the temperature of the wire increased its electrical resistance markedly, so that the measurement of the intensity of the electric current along it could indicate temperature changes of a ten-millionth of a degree.

In this way, Langley was able, for instance, to do away with the obscuring effects of absorption differences, and to show that it was the yellow portion of the spectrum which was, in fact, present in the greatest intensity and produced the greatest heat rise—as Herschel had originally assumed. (Oh, well, if Herschel had had better instruments, and if his observations had fulfilled his expectations, he would never have thought of looking outside the spectrum and he would not have discovered infrared radiation.)

Moving into the infrared region, Langley showed that there was infrared radiation over a stretch of wavelengths from the 760-nanometer length of the longest visible wavelengths of red light up to 3,000 nanometers. (1,000 nanometers, or 1,000 billionths of a meter, is equal to 1 micrometer, which is 1 millionth of a meter. Therefore, 3,000 nanometers is usually described as 3 micrometers.)

This means that the frequency of the infrared waves varies from 4.0×10^{14} (400 trillion), at the point where the visible spectrum ends, down to 1.0×10^{14} (100 trillion).

Starting with 100 trillion, we must double twice to reach 400 trillion. Therefore, attached to the one octave of visible light are two octaves of invisible infrared radiation.

The infrared spectrum seems to cut off sharply at a frequency of 100 trillion (or a wavelength of 3 micrometers), at least as far as the solar spectrum is concerned. Is that all there is, and is there no radiation of longer wavelength and lower frequency that can exist?

For that matter, what about the other end of the spectrum? If there is radiation beyond the red end, is there also radiation beyond the violet end?

We'll take up these questions and others in the next chapter.

II

Four Hundred Octaves

I have trouble describing my sense of humor, except by use of the adjective "puckish," which is derived from the description of Puck's practical jokes, in Act II, Scene i, of *A Midsummer Night's Dream*.

My dear wife, Janet, on the other hand, is inclined to use the adjective "perverted," instead, and many a time and oft I have recognized that one of my remarks has hit home because of Janet's cry of "Oh, *Isaac!*"

As a matter of fact, I hear that cry from others as well. The only person with whom I am safe in this respect is my beautiful blond-haired, blue-eyed daughter (who now lives in New Jersey, with her master's degree in social work nicely framed). She never says "Oh, *Isaac!*" She wouldn't dream of it. What she says is "Oh, *Dad!*"

Other remarks are harder to take.

One time, two of our very dearest friends were coming on a visit, and Janet, cocking an eye at the clock, said, "I do wish, Isaac, that you would take out the garbage before Phyllis and Al arrive."

"Certainly, dear," said I, all compliance. I gathered up the garbage container, opened the door, stepped out into the corridor; and there were Phyllis and Al coming toward me, big grins on their faces, arms outstretched in greeting. And there was I, laden with garbage.

I had to pass it off with some off-the-cuff witticism, so I said, "Hi! I just said to Janet that you two were about to arrive and that seemed to remind her that I ought to take out the garbage."

And two things happened. The first (quite expected) was Janet's anguished cry from within the apartment—"Oh, *Isaac!*"—ringing out simultaneously with an identical cry from Phyllis.

The second was Al's jovial laugh, as he said, "Don't worry about it, Janet. Nobody takes Isaac seriously."

Imagine! Here I go to enormous trouble writing very serious essays in every issue of *Fantasy and Science Fiction,* and he attacks my credibility just because of my bubbling, irrepressible jocosity.

Fortunately, I know that all my Gentle Readers take me seriously indeed, and so I will continue with the subject broached in the previous chapter.

In the previous chapter, I talked about the spectrum of visible light, and of the fact that William Herschel discovered, in 1800, that there was invisible light beyond the red end of the spectrum, light we now call "infrared radiation." The solar spectrum contains one octave of visible light, stretching from a frequency of 800 trillion cycles per second at the shortest-wave violet, to one of 400 trillion cycles per second at the longest-wave red. Beyond the red in the solar spectrum are two octaves of infrared radiation, extending down to a frequency of 100 trillion cycles per second.

But if there is something beyond the red, might there not be something beyond the violet as well?

That part of the story begins in 1614, when an Italian chemist, Angelo Sala (1576–1637), reported that silver nitrate, a perfectly white compound, darkened on exposure to the Sun.

This happens to other silver compounds as well, and nowadays we know what happens. Silver is not a very active element and it does not hold on to other atoms particularly tightly. The molecules of a compound, such as silver nitrate or silver chloride, can easily be broken apart, and when that happens very fine granules of metallic silver are deposited here and there among the tiny crystals of the compound. Finely divided silver happens to be black, so the compound darkens.

Light waves radiated by the Sun contain enough energy to split the molecules of silver compounds and so light will darken them. This sort of thing is an example of a "photochemical reaction."

About 1770, the Swedish chemist Carl Wilhelm Scheele (1742–86) studied the effect of sunlight on silver compounds, and he had the solar spectrum available to him (which Sala had not had). Scheele soaked thin strips of paper in solutions of silver nitrate and placed them in different parts of the spectrum. It was clear that the colors were more effective in darkening the compound as he approached the violet end of the spectrum.

This is no surprise today, of course, since we know that the energy of light goes up with frequency. Naturally, the higher the energy of a particular type of light, the greater the likelihood of that type of light breaking the chemical bonds within a molecule.

But then, in 1800, Herschel discovered infrared radiation. It occurred to a German chemist, Johann Wilhelm Ritter (1776–1810), that there might well be something beyond the other end of the spectrum, and he set about checking the matter.

In 1801, he soaked strips of paper in a solution of silver nitrate, as Scheele had done thirty years before. Ritter, however, placed strips *beyond* the violet, too, in a region where no light was visible. It was with considerable satisfaction that he found, and reported, that darkening proceeded fastest in that apparently lightless region.

At first the spectral region beyond the violet was referred to as "chemical rays," because the only way it could be studied was through its photochemical properties.

Those very photochemical properties, however, led to the development of photography. Silver compounds were mixed with a gelatinous material that was then smeared over a glass plate and enclosed in a dark box. Bright light was allowed to enter the box for a short period of time and was focused on the gelatinous material by way of a lens. Wherever the light struck, there would be darkening, so that a photographic negative was produced. From this, a photographic positive could be produced that could be treated chemically so that the pattern of light and dark was permanently fixed.

Soon after the French inventor Louis J. M. Daguerre (1789–1851) produced the first barely practical photographic process in 1839, it was seized upon by scientists to record observations involving light.

In 1842, for instance, the French physicist Alexandre Edmond Becquerel (1820–91) took the first successful photograph of the solar spectrum.

The eye, as it happens, can see just those frequencies of light that produce appropriate photochemical changes in the retina; that is, light with frequencies ranging from 800 trillion to 400 trillion cycles per second. The camera, on the other hand, can detect those frequencies of light that produce chemical breakdowns and darkening in the silver compounds on the photographic plate. Since light is less energetic the shorter the frequency, the camera can barely see red light, which is easily visible to the eye, and cannot see infrared radiation at all, any more than the eye can.

Beyond the violet, however, where the frequencies are higher still and the energies greater, the silver compounds break down quickly, so that the camera can see the region beyond the violet easily, even though the human eye cannot. Becquerel succeeded in photographing the solar spectrum beyond the violet and showed quite plainly that the spectrum was a continuous structure, substantially wider than was optically visible. The region beyond the violet even contained spectral lines, exactly as the visible region did.

From then on, it became customary to speak of the region beyond the violet as consisting of "ultraviolet radiation," the prefix "ultra" being Latin for "beyond."

In 1852, the Irish physicist George Gabriel Stokes (1819–1903) discovered that quartz is far more transparent to ultraviolet radiation than ordinary glass is. He therefore constructed prisms and lenses of quartz and found he could photograph a longer stretch of ultraviolet in the solar spectrum than could be photographed through glass.

It turned out that the solar spectrum contained a stretch of ultraviolet radiation from the 400-nanometer wavelength of the shortest-wave violet down to about 300 nanometers. This amounts to just under half an octave of ultraviolet, from a frequency of 800 trillion to 1,000 trillion cycles per second.

The solar spectrum contained, therefore, one octave of visible light, sandwiched between two octaves of infrared radiation, and not quite one half octave of ultraviolet radiation.

The absence of any farther stretch of ultraviolet turned out to be a good thing. Light produces photochemical changes in the skin, and does so the more effectively as the frequency increases. Visible light does little, but ultraviolet darkens the skin by stimulating the production of the dark pigment, melanin. If a particular skin is fair and isn't much good at producing melanin (mine, for instance), it reddens and burns instead. If the solar spectrum extended beyond the 1,000 trillion per second mark in frequency, the changes in living tissue would be more extensive and might actually preclude the existence of life that was exposed to sunlight.

The solar spectrum, then, includes radiation in the frequency range from 1,000 trillion cycles per second for the shortest ultraviolet to 100 trillion per second for the longest infrared. There are three questions we might now ask:

1. Is that all there is? Is it impossible for there to be radiation of higher frequencies than 1,000 trillion, or lower than 100 trillion, cycles per second?

2. If higher and lower are indeed possible, why do they not show up in the solar spectrum? Is the Sun incapable of producing those very high and very low frequencies or are they produced but, for some reason, fail to reach us?

3. And if there are very high and very low frequencies possible, then how high and how low? Are there any limits at all?

The first question was quickly answered, since scientists had no trouble producing ultraviolet radiation that was higher-frequency and infrared radiation that was lower-frequency than anything in the solar spectrum.

Stokes himself used an electric spark as a source of high-frequency radiation. The sparks emitted light that was richer in ultraviolet than sunlight was, and higher-frequency ultraviolet, too.

Stokes and other physicists of his time were able to follow ultraviolet down to a wavelength of 200 nanometers, which is equivalent to a frequency of about 1,500 trillion per second. That gave them just about a full octave of ultraviolet.

In the twentieth century, advances in photographic technology made it possible to go beyond the 200-nanometer mark in wavelength, even down to 10 nanometers. The frequency region from 800 trillion to 1,500 trillion cycles per second is sometimes called the "near ultraviolet," while the region from 1,500 trillion up to as far as 30,000 trillion cycles per second is called the "far ultraviolet."

Where infrared radiation was concerned, it became possible to observe and study low-energy radiation emitted by heated bodies that produced frequencies of infrared radiation far lower than the 100 trillion per second that seemed the limit in the solar spectrum. Eventually, waves approaching 1 millimeter (that is, 1,000,000 nanometers) were observed, and 1 millimeter can be taken as the limiting wavelength of infrared. This represents a frequency of 0.3 trillion (or 300 billion) cycles per second.

The spectrum would seem to stretch, then, from frequencies as little as 0.3 trillion to as much as 30,000 trillion cycles per second in frequency (or from 3×10^{11} / second to 3×10^{16} / second). This is a total range of over 16 octaves. Of these, 5 octaves are ultraviolet radiation, 1 octave is visible light, and 10 are infrared radiation. The invisible light outweighs the visible by a factor of 15.

Now to the second question. Why is the solar spectrum more limited in both directions than is the radiation which can be studied in the laboratories? Scientists didn't really think the solar spectrum was as limited as it appeared to be, and investigation of the upper atmosphere during the early twentieth century made it clear they were right.

The atmosphere is opaque to most radiation outside the visible octave. Ozone, in which the upper atmosphere is rich, blocks the shorter range of ultraviolet radiation. The longer range of infrared radiation is absorbed by various atmospheric components such as carbon dioxide and water vapor.

If sunlight could be studied outside the atmospheric blanket of Earth, it would surely be found to have a spectrum that included the full range of ultraviolet and infrared radiation, and probably beyond that on either side. By mid-twentieth century, sunlight *was* so studied and was found indeed to be very wide-spectrum.

That brings us to the third question. Are there absolute limits to radiation in either direction? Are there radiations with a longest possible and shortest possible wavelength, or (the equivalent) a lowest possible and highest possible frequency?

An approach to an answer to that originated in the study of electricity and magnetism.

These were originally thought to be two independent phenomena, but, in 1820, the Danish physicist Hans Christian Oersted (1777–1851) discovered, rather by accident, that an electric current produced a magnetic field that could affect the needle of a magnetic compass.

Other physicists immediately began to investigate this surprising state of affairs and it was quickly found that if a conductor cut through the lines of force of a magnetic field, a current of electricity could be induced in that conductor (the foundation of our modern electrified society).

In fact, the further research went, the more intimately electricity and magnetism seemed to be related. It began to be suspected that one could not exist without the other—that there was not an electric field and a magnetic field, but a combined "electromagnetic field."

In 1864, the Scottish mathematician James Clerk Maxwell (1831–79) devised a set of four comparatively simple equations that described, with surprising accuracy, the full behavior of electromagnetic phenomena, and these set the notion of the electromagnetic field on a lastingly firm foundation.

Thus, the two great physical revolutions of the twentieth century, relativity and quanta, modified nearly everything in classical physics, even Isaac Newton's theory of gravitation—but they left Maxwell's equations untouched.

The most unexpected result of the equations was that Maxwell was able to show that an electric field of changing intensity had to produce a magnetic field of changing intensity, which in turn produced an electric field of changing intensity and so on. The two effects leapfrogged, so to speak, and produced a radiation that had the properties of a transverse wave and spread outward in all directions equally. It was like dropping a pebble on the surface of a still pond, setting up a series of ripples spreading outward in all directions from the center of disturbance.

In the case of an electromagnetic field, the result is "electromagnetic radiation."

Maxwell was able to work out the speed of propagation of such electromagnetic radiation from his equations. It turned out to be equal to the ratio of certain values in his equations and this ratio proved to be 300,000,000 meters per second.

This was precisely the speed of light, which also had the properties of a transverse wave. Maxwell could not believe that this was a coincidence. He assumed that light was an example of an electromagnetic radiation, and that its varying wavelengths depended on varying rates at which electromagnetic fields were oscillating.

What electromagnetic fields?

Maxwell couldn't say, but his equations worked and he was convinced the fields were there. It was not till well after his premature death that he was shown to be completely correct in this respect.

We now know that the atom consists of subatomic particles, two of which, the electron and the proton, are electrically charged. They give rise to oscillating electromagnetic fields.

If we look at it in what modern physicists would call an unsophisticated manner, we could imagine electrons revolving about atomic nuclei, planet-fashion, and thus oscillating from one side of the nucleus to the other hundreds of trillions of times a second. The frequency of such an oscillation would equal the frequency of the light wave inevitably produced. Different frequencies would arise from electrons of different atoms, or from different electrons of the same atoms, or even from the same electrons of the same atoms under different conditions.

Instead of speaking of a light spectrum, then, we now speak of an "electromagnetic spectrum," and all the different frequencies in the spectrum reflect the different frequencies that can affect an oscillating electromagnetic field. There are therefore no fundamental distinctions between ultraviolet radiation, visible light, and infrared radiation. They represent a smooth continuum which is inevitably divided into three classes only through the accident that some frequencies, and not others, affect the chemicals in our retinas in such a way as to produce a sensation which our brains interpret as sight.

In theory, an electromagnetic field can oscillate at any frequency, so that electromagnetic radiation of any frequency can be produced. In particular, there seemed no theoretical reason why electromagnetic radiation with frequencies far lower than any of those in the infrared range might not be produced, or with frequencies far higher than any of those in the ultraviolet range.

Maxwell therefore predicted the existence of radiations outside (and even well outside) the observed limits.

This prediction was proved correct just twenty-four years later, in 1888 (something I'll take up in the next chapter). Maxwell would have been fifty-seven years old in that year, and would have viewed the discovery with great satisfaction, but he had died prematurely, nine years before, of cancer.

For the remainder of the chapter, let me speculate on reasonable limits for the electromagnetic spectrum in both directions.

As an electromagnetic field oscillates more and more slowly, the radiation produced is of lower and lower frequency and of longer and longer wavelength. If the oscillation were 300,000 cycles per second (rather than the hundreds of trillions required to produce light waves) you would have each wave one kilometer long. If the oscillation were only one cycle per second, each wave would be 300,000 kilometers long, and so on.

To be sure, as the waves grow longer and longer, their energy content decreases, and it is easy to produce waves that are so long that no present-day instrumentation could detect them. We can always assume more and more delicate instruments, however, and ask whether we would ever reach a wavelength so long and nonenergetic that no conceivable instrument would serve.

Suppose, then, we imagine an electromagnetic wave that is so long that one oscillation reaches across the full width of the Universe. Anything longer than that steps over the Universe, so to speak, and could not conceivably interact with anything in it, so that it could not be detected even in principle. We will therefore take it that the width of the Universe is the longest wavelength any significant electromagnetic radiation could have.

I usually use the figure 25,000,000,000 light-years as the diameter of the Universe. (My good friend John D. Clark, one-time science fiction writer, has recently argued that this is twice what it should be and he may be right, but let's stick with it for fun.) The rate of oscillation, producing a wavelength equal to the diameter of the Universe would then be one cycle per 25,000,000,000 years, or one cycle per 790,000,000,000,000,000 seconds. This is, roughly, 10^{-18} cycles per second.

Next, suppose we go in the other direction, and imagine wavelengths that are shorter and shorter and shorter; and, therefore, frequencies (and energies) that are higher and higher and higher.

Here, it would seem, there should be no limit. The size of the Universe might well set an upper limit to length, but what could set a lower limit?

Thanks to quantum theory, we know that the higher the frequency, the higher the energy, and we can imagine an electromagnetic wave

of frequency so high that it contains all the energy in the Universe. There can be no higher frequencies than that.

Almost all the energy in the Universe is in the form of mass. Suppose, then, we ignore the energy of the electromagnetic radiation that already exists, and the energy involved in the motions of mass. We can also ignore the possible rest-mass of the neutrinos, since this (see "Nothing and All," in *Counting the Eons*, Doubleday, 1983) is still a very iffy thing.

Therefore, we are able to make the reasonable guess that there are 100,000,000,000 galaxies in the Universe and that each galaxy possesses a mass equal to 100,000,000,000 times that of our Sun. (There are galaxies, including our own, that are considerably more massive than this; but there are also those that are considerably less massive.)

In that case, the mass of the Universe would therefore be 10,000,000,000,000,000,000,000, or 10^{22} times that of the Sun. Since the Sun's mass is just about 2×10^{30} kilograms, the mass of the Universe would be 2×10^{52} kilograms.

According to relativity theory, $e = mc^2$, where e is energy, m is mass, and c is the speed of light. According to quantum theory, $e = hf$, where h is Planck's constant, and f is frequency. (Actually, frequency is usually represented by the Greek letter "nu," but I do not wish to present problems for the Noble Printer.)

If we combine the two equations, we find that $f = mc^2/h$. Using the correct sets of units (trust me!) we can let m equal 2×10^{52}, c^2 equal 9×10^{16}, and h equal 6.6×10^{-34}. Working out the equation, we find a wavelength of 2.7×10^{102} cycles per second. The corresponding wavelength of radiation of such a frequency is 10^{-94} meters.

The total range of electromagnetic radiation, then, is from 10^{-18} cycles per second for a wave as long as the Universe is wide, to 2.7×10^{102} cycles per second for a wave so short as to contain the mass of the Universe. This is a range of 120 orders of magnitude. There are roughly 10 octaves to 3 orders of magnitude, so the full conceivable stretch of electromagnetic radiations is about 400 octaves.

Of these, there are just under 100 octaves beyond the infrared, and just under 300 octaves beyond the ultraviolet. The tiny band of ultraviolet, visible light, and infrared covers 16 octaves in between and makes up 1/25 of the whole. Visible light, at one octave, makes up 1/400 of the whole.

It seems to me that at the time of the big bang, the Universe must have made its appearance as a single particle of nearly zero size and of Universal mass. I called such a particle a "holon" in "The Crucial Asymmetry" (see *Counting the Eons*, Doubleday, 1983) but Tom Easton, in the August 1979 issue of *Analog* preceded me with a similar notion of what he called a "monobloc." Alas, I was unaware of that, and I cheerfully acknowledge his priority.

The diameter of the holon would then be 10^{-94} meters. Compare this to a proton, which has a diameter of 10^{-15} meters. The diameter of a proton is 10^{79} times that of a holon, whereas the diameter of the Universe is 10^{41} times that of a proton. To the holon, then, a proton would be far, far vaster than the whole Universe to a proton.

III

The Three Who Died Too Soon

I have just returned from the Philcon—the annual convention sponsored by the Philadelphia Science Fiction Society.

It was extremely successful, I thought. It was well-attended, efficiently run, with an excellent art show and a bustling huckster room. Joe Haldeman was the guest of honor and gave an absolute whiz-bang of a talk that was greeted with great enthusiasm by the audience. This cast me down, I fear, for I was scheduled to follow him and I had to extend myself to the full, I assure you.

But what I enjoyed the most was the costumé show that was won by a young man who had designed an unbelievably clever "satyr" costume. He carried a pipes-of-Pan about his neck, wore horns that blended perfectly with his hair, and capered about on goat legs that looked like the real thing.

My own private pleasure reached its peak, though, when three people came out on stage to the accompaniment of portentous music in order to represent "Foundation," "Foundation and Empire," and "Second Foundation," the three parts of my well-known "Foundation Trilogy." They were all three swathed in black robes and all looked somber. I watched curiously, wondering how they could possibly represent those three highly intellectual novels.

Suddenly, all three flashed—flinging open their robes and revealing themselves as three very incompletely clothed young people. The first and third were young men, in whom my interest was necessarily limited, and who were each wearing very little more than corsets (the first and second "foundation," as I at once understood).

The middle person was a young woman of pronounced beauty, both of face and figure, and she wore a corset, too. She, however, was "Foundation and Empire," and the Empire portion, I gathered, was the only other item she wore—a brassiere that did a delightfully poor job of concealing what it was meant to support.

After a few moments of surprise and enchantment, my scientific training asserted itself. If careful observation is required, it must be made under the most favorable conditions. I therefore stood up and leaned forward.

Whereupon, from near me, a voice could be heard saying, "That's five bucks you owe me. He stood up."

That was a sensationally easy bet to win—and another sensationally easy bet to win is that I will now proceed with still a third chapter on the electromagnetic spectrum.

In the last two chapters, I discussed visible light, infrared radiation, and ultraviolet radiation. The frequencies in question ranged from as little as 0.3 trillion cycles per second for the lowest-frequency infrared to as much as 30,000 trillion cycles per second for the highest-frequency ultraviolet.

In 1864, however, as I said, James Clerk Maxwell had evolved a theory that made it seem that such radiations arose from an oscillating electromagnetic field (hence, "electromagnetic radiation") and that the frequency could be any value from much higher than 30,000 trillion cycles per second to much lower than 0.3 trillion cycles per second.

A good, airtight, well-thought-out theory is a delight, but it becomes even more delightful if some phenomenon, which has never been observed, is predicted by the theory—and is then observed. The theory points, and you look, and, behold! it's there. The chances of doing so, however, do not seem great.

It is possible to make an electric current (and hence an electromagnetic field) oscillate. Such oscillations are comparatively slow, however, and if, as is predicted by Maxwell's equations, they produce an electromagnetic radiation, the frequency is far lower than even the lowest-frequency infrared radiation. Millions of times lower. Surely, the detection methods that worked for the familiar radiations in the region of light and its immediate neighbors would not work for something so far removed in properties.

Yet detected it would have to be—and in such detail that the waves could be shown to have the nature and properties of light.

Actually, the thought of oscillating electric currents producing some sort of radiation antedated Maxwell.

The American physicist Joseph Henry (1797–1878) had discovered the principle of "self-induction" in 1832 (I won't go into that or I'll never get through the ground I want to cover in this essay). In 1842, he tackled certain confusing observations that made it seem uncertain, in some cases, in which direction an electric current was moving. Under certain conditions, in fact, it seemed to be moving in both directions.

Henry, using his self-induction principle, reasoned that when a Leyden jar (or a capacitor, generally) is discharged, for instance, it overshoots the mark, so that a current flows out, then finds it must flow back, overshoots the mark again, flows in the first direction and so on. In short, the electric current oscillates much as a spring might. What's more, it can be a damped oscillation, such that each overshooting of the mark is less than the one before until the current flow settles down to zero.

Henry knew that a current flow produced an effect at a distance (it would make the needle of a distant magnetic compass veer, for instance) and felt that this effect would change and shift with the oscillations so that one would have a wave-like radiation issuing out from the oscillating current. He even compared the radiation to light.

This was just a vague speculation with Henry, but it is a distinguishing mark of great scientists that even their vague speculations have an uncanny habit of being right. Nevertheless, it was Maxwell, a quarter century later, who reduced the whole matter to a clear mathematical statement and it is he who deserves the credit.

Not all scientists accepted Maxwell's reasoning, however. One who didn't was the Irish physicist George Francis FitzGerald (1851–1901), who wrote a paper categorically maintaining that it was impossible for oscillating electric currents to produce wave-like radiations. (FitzGerald is very well known by name to science fiction readers, or should be, since it was he who originated the concept of "the FitzGerald contraction.")

It was quite possible that scientists might choose up sides, some following Maxwell and some FitzGerald, and argue over the matter forever, unless the electric oscillation waves were actually detected, or unless some observation were made that clearly showed such waves to be impossible.

It's not surprising, then, that Maxwell would feel keenly the importance of detecting these very low-frequency waves. It was with dejection that he felt locating them was so difficult as to be next door to impossible.

And then, in 1888, a thirty-one-year-old German physicist, Heinrich Rudolph Hertz (1857–94), managed to do the job and to establish Maxwell's theory on a firm observational foundation. Had Maxwell lived, his pleasure at seeing that establishment would have been outdistanced, I am sure, by his surprise at seeing how easy the detection was and how simply it was managed.

All Hertz needed was a rectangular wire, with one side adjustable so that it could be moved in and out, and the opposite side possessing a small gap. The wire at each side of the gap ended in a small brass knob. If a current were somehow started in that rectangular wire, it could leap the gap, producing a small spark.

Hertz then set up an oscillating current by discharging a Leyden jar. If it produced electromagnetic waves, as Maxwell's equations predicted, those waves would induce an electric current in Hertz's rectangular detector (to which no other source of electricity was attached, of course). A spark would then be produced across the gap, and this would be visible evidence of the induced electric current, and, therefore, of the waves that did the inducing.

Hertz got his sparks.

By moving his receiver about in different directions and at different distances from the oscillating current that was the source of the waves, Hertz found the sparks growing more intense in places and less intense in others as the waves were at higher or lower amplitude. He could, in this way, map out the waves, determine the wavelength, and show that they could be reflected, refracted, and made to exhibit interference

phenomena. He could even detect both electric and magnetic properties. In short, he found the wave entirely similar to light, except for
their wavelengths, which were in the meter range rather than the micrometer range. Maxwell's electromagnetic theory was well and truly
demonstrated, nine years after Maxwell's death.

The new waves and their properties were quickly confirmed by other
observers and they were termed "Hertzian waves."

Neither Hertz nor any of those who confirmed his findings saw the
discovery as of any importance other than as the demonstration of the
truth of an elegant scientific theory.

In 1892, however, the English physicist William Crookes (1832–
1919) suggested that Hertzian waves might be used for communication. They moved in straight lines at the speed of light, but were so
long-wave that objects of ordinary size were simply not opaque to
them. The long waves moved around and through obstacles. The waves
were easily detected and, if they could be started and stopped in a
careful pattern, they could produce the dots and dashes of the tele-
graphic Morse code—and without the need of the complicated and
expensive system of thousands of kilometers of copper wires and re-
lays. Crookes was, in short, suggesting the possibility of "wireless
telegraphy."

The idea must have sounded like "science fiction" (in the pejora-
tive sense used by ignorant snobs), and Hertz, alas, did not see it
come true. He died in 1894, at the age of forty-two, of a chronic
infection that, these days, would probably have been easily cured by
antibiotics.

Only months after Hertz's death, however, an Italian engineer,
Guglielmo Marconi (1874–1937), then only twenty years old, read of
Hertz's findings and instantly got the same idea Crookes had had.

Marconi used the same system for producing Hertzian waves that
Hertz himself had used, but set up a much improved detector, a so-
called coherer. This consisted of a container of loosely packed metal
filings, which ordinarily conducted little current, but conducted quite
a bit when Hertzian waves fell upon it.

Gradually, Marconi improved his instruments, grounding both the
transmitter and receiver. He also used a wire, insulated from the earth,
which served as an antenna or aerial to facilitate both sending and
receiving.

He sent signals across greater and greater distances. In 1895, he sent a signal from his house to his garden and, later, across a distance of over a kilometer. In 1896, when the Italian Government showed itself uninterested in his work, he went to England (his mother was Irish and Marconi could speak English) and sent a signal across a distance of fourteen kilometers. He then applied for and received the first patent in the history of wireless telegraphy.

In 1897, again in Italy, he sent a signal from land to a warship twenty kilometers away, and in 1898 (back in England) he sent a signal across a distance of thirty kilometers.

He was beginning to make his system known. The seventy-four-year-old British physicist Lord Kelvin paid to send a "Marconigram" to his friend the British physicist G. G. Stokes, then seventy-nine years old. This communication between two aged scientists was the first commercial message by wireless telegraphy. Marconi also used his signals to report the yacht races at Kingstown Regatta that year.

In 1901, Marconi approached the climax. His experiments had already convinced him that Hertzian waves followed the curve of the Earth instead of radiating straight outward into space as electromagnetic waves might be expected to do. (It was eventually found that Hertzian waves were reflected by the charged particles in the "ionosphere," a region of the upper atmosphere. They traveled around the Earth's bulge by bouncing back and forth between ground and ionosphere.)

He made elaborate preparations, therefore, to send a Hertzian-wave signal from the southwest tip of England across the Atlantic to New-foundland, using balloons to lift the antennae as high as possible. On December 12, 1901, he succeeded.

To the British, the technique has remained "wireless telegraphy" and the phrase is usually shortened to "wireless."

In the United States, the technique was called "radio-telegraphy," meaning that the key carrier of the signal was an electromagnetic radiation rather than a current-carrying wire. For short, the technique was called "radio."

Since Marconi's technique made headway fastest in the United States, which was by now the most advanced nation in the world from the technological standpoint, "radio" won out over "wireless." The world, generally, speaks of radio, now, and December 12, 1901, is usually thought of as the day of "the invention of radio."

In fact, Hertzian waves have come to be called "radio waves" and the older name has dropped out of use. The entire portion of the electromagnetic spectrum from a wavelength of one millimeter (the upper boundary of the infrared region) to a maximum wavelength equal to the diameter of the Universe—a stretch of 100 octaves—is included in the radio-wave region.

The radio waves used for ordinary radio transmission have wavelengths of from about 190 to 5,700 meters. The frequency of these radio waves is therefore from 530,000 to 1,600,000 cycles per second (or from 530 to 1,600 kilocycles per second). A "cycle per second" is now referred to as a "hertz" in honor of the scientist, so we might say that the frequency range is from 530 to 1,600 "kilohertz."

Higher-frequency radio waves are used in FM, still higher-frequency in television.

As years went by, radio came into more and more common use. Methods for converting radio signals into sound waves were developed so that you could hear speech and music on radio, and not just the Morse code.

This meant that radio could be combined with ordinary telephonic communication to produce "radio-telephony." In other words, you could use the telephone to communicate with someone on a ship in mid-ocean, when you yourself were in mid-continent. Ordinary telephone wires would carry the message across land, while radio waves would carry it across the sea.

There was a catch, however. Wire-conducted electricity could produce sound that was clear as a bell (Alexander Graham, of course), but air-conducted radio waves were constantly being interfered with by the random noise we call "static" (because one cause is the accumulation of a static electrical charge upon the antenna).

Bell Telephone was naturally interested in minimizing static, but in order to do that, they had to learn as much as possible about the causes of it. They assigned the task of doing so to a young engineer named Karl Guthe Jansky (1905–50).

One of the sources of static was certainly thunderstorms, so one of the things that Jansky did was to set up a complicated aerial, consisting of numerous rods, both vertical and horizontal, which could receive from different directions. What is more, he set it up on an automobile frame equipped with wheels, so that he could turn it this way and that in order to tune in on any static he did detect.

Using this device, Jansky had no trouble detecting distant thunderstorms as crackling static.

It was not all he got, however. While he was scanning the sky, he also got a hissing sound quite different from thunderstorm crackles. He was clearly getting radio waves from the sky, radio waves that were generated neither by human beings nor thunderstorms. What's more, as he studied this hiss from day to day, it seemed to him that it was not coming from the sky generally but, for the most part, from some particular part of it. By moving his aerial system properly he could point it in a direction from which the sound was most intense—and this spot moved across the sky, rather as the Sun did.

At first, it seemed to Jansky that the radio-wave source *was* the Sun, and if the Sun had happened to be at a high sunspot level at the time, he would have been right.

However, the Sun was at low activity at the time and what radio waves it emitted could not be detected by Jansky's crude apparatus. That, perhaps, was a good thing, for it turned out that Jansky was onto something bigger. At the start, his apparatus did indeed seem to be pointing toward the Sun when it was receiving the hiss at maximum intensity, but day by day Jansky found his apparatus pointing farther and farther away from the Sun.

The point from which the hiss was originating was fixed with respect to the stars, while the Sun was not (as viewed from Earth). By the spring of 1932, Jansky was quite certain that the hiss was coming from the constellation of Sagittarius. It was only because the Sun was in Sagittarius when Jansky detected the cosmic hiss that he initially confused the two.

The center of the Galaxy happens to be in the direction of Sagittarius, and what Jansky had done was to detect the radio emissions from that center. The sound came to be called the "cosmic hiss" because of this.

Jansky published his account in the December 1932 issue of *Proceedings of the Institute of Radio Engineers* and that marks the birth of "radio astronomy."

But how could radio waves reach Earth's surface from outer space when they were reflected by the ionosphere? The ionosphere keeps radio waves originating on Earth from moving out into space, and should keep those originating in space from moving down to Earth's surface.

It turned out that a stretch of about eleven octaves of the very shortest radio waves (called "microwaves"), just beyond the infrared, were not reflected by the ionosphere. These very short radio waves could move right through the ionosphere, either from Earth into space, or from space down to Earth. This stretch of octaves is known as the "microwave window."

The microwave window encompasses radiation with wavelengths from about 10 millimeters to about 10 meters, and frequencies from 30,000,000 cycles per second (30 megahertz) to 30,000,000,000 cycles per second (30,000 megahertz).

Jansky's apparatus happened to be sensitive to a frequency just inside the lower limit of the microwave window. A little bit lower and he might not have detected the cosmic hiss.

The news of Jansky's discovery made the front page of the New York *Times,* and justifiably so. With the wisdom of hindsight, we can at once see the importance of the microwave window. For one thing, it included seven octaves as compared to the single octave of visible light (plus a bit extra in the neighboring ultraviolet and infrared). For another, light is useful for nonsolar astronomy only on clear nights, whereas microwaves would reach Earth whether the sky was cloudy or not, and for that matter they could be worked with in the daytime as well for the Sun would not obscure them.

Nevertheless, professional astronomers paid little attention. The astronomer Fred Lawrence Whipple (1906–   ), who had just joined the Harvard faculty, did discuss the matter with animation, but he had the advantage of being a science fiction reader.

We can't blame astronomers too much, however. After all, there was nothing much they could do about it. The instrumentation required for receiving microwaves with sufficient delicacy to be of use in astronomy simply didn't exist.

Jansky himself didn't follow up his discovery. He had other things to do, and his health was not good. He died of a heart ailment at the age of forty-three and barely lived to see radio astronomy begin to stir. By a strange fatality, then, three of the key scientists in the history of radio, Maxwell, Hertz, and Jansky, each died in his forties and did not live to see the true consequences of his work, even though each would have done so if he had lived but another decade.

Still, radio astronomy was not entirely neglected. One person, an amateur, carried on. This was Grote Reber (1911–   ), who had become an enthusiastic radio ham at the age of fifteen. While he was still a student at the Illinois Institute of Technology, he took Jansky's discovery to heart and tried to follow. For instance, he tried to bounce radio signals off the Moon and detect the echo. (He failed, but the idea was a good one, and, a decade later, the Army Signal Corps, with far more equipment at its disposal, was to succeed.)

In 1937, Reber built the first radio telescope in his back yard in Wheaton, Illinois. The reflector, which received the radio waves, was 9.5 meters in diameter. It was designed as a paraboloid so that it concentrated the waves it received at the detector at the focus.

In 1938, he began to receive and, for several years, he was the only radio astronomer in the world. He discovered places in the sky that emitted stronger-than-background radio waves. Such "radio stars," he found, did not coincide with any of the visible stars. (Some of Reber's radio stars were eventually identified with distant galaxies.)

Reber published his findings in 1942, and by then there was a startling change in the attitude of scientists toward radio astronomy.

A Scottish physicist, Robert Watson-Watt (1892–1973), had grown interested in the manner in which radio waves were reflected. It occurred to him that radio waves might be reflected by an obstacle, and the reflection detected. From the time lapse between emission and detection of reflection the distance of the obstacle could be determined and, of course, the direction from which the reflection was received would give the direction of the obstacle.

The shorter the radio waves, the more easily they would be reflected by ordinary obstacles; but if they were too short, they would not penetrate clouds, fog, and dust. Frequencies were needed that were high enough to be penetrating and yet low enough to be efficiently reflected by objects you wanted to detect. The microwave range was just suitable for the purpose, and, as early as 1919, Watson-Watt had already taken out a patent in connection with radio location by means of short radio waves.

The principle is simple, but the difficulty lies in developing instruments capable of sending out and receiving microwaves with the requisite efficiency and delicacy. By 1935, Watson-Watt had patented improvements that made it possible to follow an airplane by the radio-wave reflections it sent back. The system was called "radio detection and ranging" (to "get a range" on an object is to determine its distance). This was abbreviated to "ra. d. a. r." or "radar."

Research was continued in secrecy, and, by the fall of 1938, radar stations were in operation on the British coast. In 1940, the German Air Force was attacking those stations, but Hitler, in a fury over a minor bombing of Berlin by the RAF, ordered German planes to concentrate on London. They ignored the radar stations thereafter (not quite grasping their abilities) and found themselves consistently unable to achieve surprise. In consequence, Germany lost the Battle of Britain, and the war. With all due respect to the valor of British airmen, it was radar that won the Battle of Britain. (On the other hand, American radar detected incoming Japanese planes on December 7, 1941—but it was ignored.)

The same techniques that made radar possible, as it happened, could be used by astronomers to receive microwaves from the stars, and, for that matter, to send tight beams of microwaves to the Moon and other astronomical objects, and receive the reflections.

If anything was needed to sharpen astronomical appetites, it came in 1942, when all the British radar stations were simultaneously jammed. At first, it was suspected that the Germans had worked out a way of neutralizing radar, but that was not so at all.

It was the Sun! A giant flare had sprayed radio waves in Earth's direction and had flooded the radar receivers. —*Well*, if the Sun could send out such a flood of radio waves, and if the technology for studying them now existed, astronomers could barely wait till the war was over.

Once the war ended, developments came quickly. Radio astronomy flourished, radio telescopes became more delicate, new and absolutely astonishing discoveries were made. Our knowledge of the Universe underwent a mad growth of a kind that had previously taken place only in the decades following the invention of the telescope.

But that goes beyond the present limits of discussion. In the next chapter, we'll consider the other end of the spectrum, the portion beyond the ultraviolet, and with that, our four-chapter investigation of electromagnetic radiation will be completed.

IV

X Stands for Unknown

When one is approaching early middle age (as I have been doing for decades) it becomes necessary to make periodic visits to a periodontist. He is the fellow (in case you don't know) who tells you that your teeth are in perfect shape and as strong as steel, but that if you don't do something about your gums, all your teeth will fall out tomorrow.

He then does something about your gums, but the giveaway comes when he approaches with the anesthetic—about two quarts of it.

My periodontist has a grandmother who (he says) calls him "Golden Fingers." I, myself, prefer to refer to him, affectionately, as "the Butcher."

On a recent visit, I said to my periodontist severely, "Last time you told me to see my regular dentist because you thought some of my fillings were getting old and deteriorating, and so I did, and he promptly chained me to the chair, capped two teeth, charged me a thousand dollars—and God's going to get you for that."

"He already did," said the villain, calmly. "You're back!"

—Well, I *am* back, and with the fourth chapter of the story of the electromagnetic spectrum.

In the previous chapter, I talked about radio waves, that region of long-wave, low-frequency electromagnetic waves beyond the infrared. These were discovered by Hertz in 1888 and, by that discovery, the usefulness and validity of the Maxwell equations were amply demonstrated.

By those same equations, if there were electromagnetic waves beyond, and even far beyond, the infrared, there should equally well be electromagnetic waves beyond, and even far beyond, the ultraviolet.

Nobody was looking for them, however.

What did exercise the interest of many physicists in the 1890s were the "cathode rays." These were a form of radiation that streamed across an evacuated cylinder from a negative electrode ("cathode") sealed inside, once an electric circuit was closed.

The study reached its climax in 1897, when an English physicist, Joseph John Thomson (1856–1940), demonstrated, quite conclusively, that cathode rays did not consist of waves, but of a stream of speeding particles.* What was more (much more), those particles were far less massive than even the least massive atoms. The mass of the cathode-ray particle was only 1/1837th that of a hydrogen atom, and Thomson called it an "electron." He received the Nobel Prize for physics for this in 1906.

The electron was the first subatomic particle discovered and it was one of a series of discoveries in the 1890s that totally revolutionized physics.

It was not, however, the first of these discoveries. The first to initiate the new age was a German physicist, Wilhelm Conrad Roentgen (1845–1923). In 1895, he was fifty years old, and the head of the department of physics at the University of Würzburg in Bavaria. He had done solid work, had published forty-eight competent papers, but was far from immortality and would undoubtedly never have risen beyond the second rank but for the events of November 5, 1895.

He was working on cathode rays, and he was particularly interested in the way in which cathode rays caused certain compounds to glow, or luminesce, upon impingement. One of the compounds that luminesced was barium platinocyanide, and Roentgen had sheets of paper coated with that compound in his laboratory.

---

* Actually, every particle has its wave aspects, and every wave has its particle aspects, and, as in the case of so many dualities in nature, you can't have one without the other. This was not understood in 1897, however.

The luminescence was very faint and, in order to observe it as well as possible, Roentgen darkened the room and enclosed the experimental apparatus within sheets of black cardboard. He could then peer into an enclosure that was totally dark, and when he turned on the electric current, the cathode rays would pass along the tube, penetrate the thin far wall, impinge upon the chemical-coated paper, and start a luminescence he would be able to see and study.

On that November 5, he turned on the current and, as he did so, a dim flash of light that was *not* inside the apparatus caught the corner of his eye. He looked up and there, quite a distance from the apparatus, was one of those sheets, coated with barium platinocyanide, and it was luminescing briskly.

He turned off the current; the coated paper darkened. He turned on the current; the coated paper gleamed again.

He took the paper into the next room and pulled down the blinds in order to darken that room too. He returned to the room with the cathode-ray tube and turned on the electricity. He walked into the next room, and closed the door behind him. The coated paper was glowing with a wall and a door between itself and the cathode-ray tube. It glowed only when the apparatus in the next room was working.

It seemed to Roentgen that the cathode-ray tube was producing a penetrating radiation that no one had reported before.

Roentgen spent seven weeks exploring the penetrative power of this radiation: what it could penetrate; what thickness of what material would finally stop it, and so on. (Later on, when he was asked what he thought when he made his discovery, he answered sharply, "I didn't think; I experimented.")

He must have been a trial to his wife during this period. He came to dinner late and in a savage mood, didn't talk, bolted his food, and raced back to the laboratory.

On December 28, 1895, he finally published his first report on the subject. He knew what the radiation *did*, but he didn't know what it *was*. Mindful of the fact that in mathematics, x is usually used to signify an unknown quantity, he called the radiation "X rays."

An alternate name at first was "Roentgen rays" in his honor, but the Teutonic "oe" is a vowel that Germans can pronounce with ease, but that is liable to break the teeth of anyone else trying to pronounce it. Consequently, X rays is what the radiation is called even today, although its nature is no longer a mystery.

It was instantly recognized that X rays could serve as a medical tool. Only four days after news of Roentgen's discovery reached America, X rays were used to locate a bullet in someone's leg. (It took a few tragic years to discover that X rays were also dangerous, and that they could cause cancer.)

In the world of science, X rays at once took up the attention of almost every physicist and this led to a rash of other discoveries, not the least of which was the discovery of radioactivity in 1896. Within a year of Roentgen's discovery, a thousand papers on X rays were published, and when, in 1901, the Nobel Prizes were first instituted, Roentgen was honored with the first Nobel Prize for physics.

X rays also made an impact on the general public. Panicky members of the New Jersey legislature tried to push through a law preventing the use of X rays in opera glasses in order to protect maidenly modesty—which is about par for the scientific literacy of elected officials.

The King of Bavaria offered Roentgen a title, but the physicist refused, recognizing quite well where the true honor of science lay. He also refused to make any attempt to patent any aspect of X-ray production or to make any financial gain from it, feeling that it was not right to do so. His reward was that he died, totally penniless, in 1923, having been impoverished by the ruinous postwar inflation in Germany.

What exactly were X rays? Some thought they consisted of streaming particles, like cathode rays. Some, including Roentgen, thought they consisted of waves, but longitudinal waves like those of sound and so were not electromagnetic. And some thought they *were* electromagnetic waves that were shorter than ultraviolet.

If X rays were electromagnetic in nature (the alternative which grew steadily in popularity), they ought to show some of the properties of other electromagnetic radiation. They should show interference phenomena.

These could be demonstrated by diffraction gratings: a transparent sheet of matter on which opaque lines are scratched at regular intervals. Radiation passing through such a grating would produce interference patterns.

The trouble was that the smaller the wavelength of the radiation, the more closely spaced the gratings had to be to produce results, and if X rays were composed of waves much smaller than those of ultraviolet, there was no technique known that would rule the gratings closely enough.

And then a German physicist, Max Theodor Felix von Laue (1879–1960), had one of those simple ideas that are blindingly brilliant. Why bother trying to scratch an impossibly fine grating when nature has done the job for you?

In crystals, the various atoms composing the substance are lined up in rows and files with rigid regularity. This is what makes the substance a crystal, in fact, and this had been known for a century. The rows of atoms correspond to the scratches on the diffraction grating and the space between to the transparent material. As it happens, the distance between atoms was just about the wavelength physicists guessed the X rays might have. Why not, then, send X rays through crystals and see what happened?

In 1912, the experiment was tried under Laue's direction, and it worked perfectly. The X rays, passing through a crystal before impinging on a photographic plate, were diffracted, and produced a regular pattern of spots. They behaved exactly as electromagnetic waves of very short wavelength would be expected to do. That settled the nature of X rays once and for all, and "X" became inappropriate (but was kept on to the present day, anyhow).

As for Laue, he was awarded the Nobel Prize in physics in 1914 for this work.

There was more to this than the mere demonstration of X-ray diffraction. Suppose a crystal of known structure was used, one in which the separation between the rows and files of atoms could be determined with reasonable precision by some method. In that case, from the details of the diffraction, the precise wavelength of the X rays being used could be determined.

In reverse, once you had a beam of X rays of known wavelength, you could bombard a crystal of unknown structural detail and, from the nature of the diffraction pattern, you could determine the location and spacing of the atoms making up the crystal.

The Australian-English physicist William Lawrence Bragg (1890–1971) was a student at Cambridge when he read of Laue's work, and he thought of the implication at once. He got in touch with his father, William Henry Bragg (1862–1942), who was a professor at the University of Leeds, and who had also grown interested in Laue's work.

Together, they worked out the mathematics and ran the necessary experiments, which worked perfectly. The results were published in 1915, and within months the two shared the Nobel Prize for physics for that year. The younger Bragg was only twenty-five when he received the prize, and he was the youngest ever to get one. He lived to celebrate the fifty-fifth anniversary of the prize, also a record.

The wavelength of X rays extends from the boundary of the ultraviolet at 10 nanometers (10^{-8} meters) down to 10 picometers (10^{-11} meters). In frequencies, X rays run from 3×10^{16} to 3×10^{19} cycles per second. That's about 10 octaves.

The distance between planes of atoms in a crystal of salt is 2.81×10^{-10} meters, and the width of an atom is about 10^{-10} meters, so you can see that X-ray wavelengths are just about in the atomic range. No wonder crystal diffraction works for X rays.

As I said earlier in the essay, the discovery of X rays led directly to the discovery of radioactivity the next year.[†]

Radioactivity involves (as the very name of the phenomenon indicates) the production of radiation. This radiation proved, like X rays, to be penetrating. Were radioactive radiations identical with X rays, then, or were they at least something similar?

In 1899, the French physicist Antoine Henri Becquerel (1852–1908), who had discovered radioactivity, found that the radioactive radiations could be deflected by a magnetic field in the same direction as cathode rays were.

That showed at once that the radioactive radiations could not be electromagnetic in nature, since electromagnetic radiations did not respond to a magnetic field at all.

Almost immediately afterward, and independently, the New Zealand-born physicist Ernest Rutherford (1871–1937) also noted the ability of a magnetic field to deflect radioactive radiations. His observations were more detailed, however. He noted that there were at least two different kinds of radioactive radiations, one of which was deflected in the manner Becquerel had noted, but the other of which was deflected in the opposite direction.

Since the cathode rays consist of negatively charged particles, it was

[†] For details, see "The Useless Metal" in *The Sun Shines Bright*, Doubleday, 1981.

clear that the radioactive radiation that was deflected in the same direction also consisted of negatively charged particles. The radioactive radiation deflected in the other direction must consist of positively charged particles.

Rutherford called the positively charged radiation "alpha rays" after the first letter in the Greek alphabet, and the other he called "beta rays" after the second letter. The names are still used to this day. The speeding particles of which these rays are composed are called, respectively, "alpha particles" and "beta particles."

During the year 1900, Becquerel, Rutherford, and the husband and wife team of the Curies, Pierre (1859–1906) and Marie (1867–1934), all worked on the radioactive radiations. All showed that the beta rays were about 100 times as penetrating as the alpha rays. (Becquerel and the Curies shared the Nobel Prize in physics in 1903, and Rutherford got his—in chemistry, to his snobbish disappointment—in 1908.)

The negatively charged beta rays were deflected to such a degree that they had to be composed of very light particles and, in that way, too, closely resembled the cathode-ray particles. Indeed, when Becquerel, in 1900, calculated the mass of the beta particles from their speed, the amount of their deflection, and the strength of the magnetic field, it became clear that beta particles did not merely resemble cathode-ray particles, but were identical to them. In brief, beta particles were electrons, and beta rays were composed of streams of speeding electrons.

This discovery made it plain that electrons were found not only in electric currents (which was all that the research with cathode rays indicated) but also in atoms that apparently had nothing to do with electricity. This was the first hint that atoms had a complicated structure, and at once physicists began to try to explain how atoms could contain electrically charged electrons and still remain electrically neutral.

As for the alpha rays, they were deflected very little by a magnetic field of an intensity that deflected beta rays a great deal. That meant that alpha rays were much more massive than electrons.

In 1903, Rutherford was able to show that alpha particles were as massive as atoms, and by 1906, he had refined his measurements to the point where he could demonstrate that they were as massive as helium atoms in particular. In 1909, in fact, he showed that alpha particles turned into helium atoms upon standing.

It was Rutherford who then went on, in 1911, to work out the concept of the nuclear atom. Each atom, he maintained, consisted of negatively charged electrons on the outskirts surrounding a very small positively charged "nucleus" at the core. Together, electrons and nucleus balanced charges and produced a neutral atom. What's more, the new concept made it clear that alpha particles were helium nuclei.

As it happened, though, alpha rays and beta rays were not the only radiations produced by radioactivity.

There was a third type of radiation, one that was discovered in 1900 by the French physicist Paul Ulrich Villard (1860–1934). He noted that some of the radiation was not deflected by the magnetic field *at all*. That radiation inevitably received the name of "gamma rays" from the third letter of the Greek alphabet.

The reason it took a while to notice the gamma rays was this:

Alpha particles and beta particles, both carrying electric charges, attracted, or repelled, electrons out of atoms, leaving positively charged ions behind. (This was only thoroughly understood after the nuclear atom was accepted.) Ions were easy to detect by the techniques of the day (and by the better techniques developed in later years). The gamma rays, which carried no electrical charge, were less efficient in forming ions and were correspondingly more difficult to detect.

The question was: What were the gamma rays?

Rutherford thought they were electromagnetic radiation that was even shorter in wavelength than X rays were. (That seemed logical since gamma rays were even more penetrating than X rays.)

The elder Bragg, however, suspected they might be high-speed particles. If they were, they would have to be electrically uncharged since they were not affected by a magnetic field. The only uncharged particles known, up to that time, were intact atoms, and they were not very penetrating. To explain the penetrating qualities of a stream of particles, they would have to be assumed to be subatomic in size, and all the subatomic particles known up to that time (electrons and atomic nuclei) were electrically charged.

If Bragg were proven correct, then, it would have been very exciting, for something entirely different—neutral subatomic particles—would have turned up. Rutherford's suggestion merely implied the same as before, only more so, for in his version gamma rays would only be "ultra-X rays."

Unfortunately, you can't force science to take the dramatic path just because you like drama. In 1914, after Laue had shown that crystals could diffract X rays, Rutherford found a crystal that would diffract gamma rays, and that settled the matter.

Gamma rays were electromagnetic in nature with wavelengths starting at the lower boundary of X rays (10^{-11} meters) and stretching down to still shorter lengths indefinitely.

A typical gamma ray had its wave more or less as long as an atomic nucleus is wide.

Dividing X rays from gamma rays at a specific wavelength is purely arbitrary. We might define them otherwise by saying that X rays are given off by changes in energy level of inner electrons, and gamma rays by changes in energy level of particles within the nucleus. It might well be, then, that some particularly energetic radiation produced by electrons might be shorter-wave than some particularly mild radiation produced by nuclei. In that case, what we call X rays and gamma rays might overlap.

That, however, is strictly a man-made problem. Two radiations of identical wavelength, one produced by electrons and the other by nuclei, are entirely identical. The wavelength is all that counts and the point of origin is of no importance, except insofar as it helps human beings indulge their passion for compartmentalization.

Are gamma rays as far as we can go in the direction of shorter and shorter wavelength?

For a time, there seemed to be a candidate for a still more energetic form of electromagnetic radiation. At least devices that could detect penetrating radiation detected *something* even when they were shielded well enough to keep off radioactive radiations. Something, therefore, was more penetrating than gamma rays.

The assumption was that this radiation came from the ground. Where else could it come from?

In 1911, an Austrian physicist, Victor Franz Hess (1883–1964), decided to confirm the obvious by taking a radiation-detecting device up in a balloon. He expected to show that when he got high enough above the ground, all signs of penetrating radiation would cease.

Not so! Instead of falling off as he rose, the penetrating radiation *increased* in intensity the higher he went. By the time he reached a height of six miles, the intensity was eight times what it had been on the ground. Hess therefore called them (in German) "high-altitude rays" and suggested they came from outer space. He received the Nobel Prize for physics in 1936 for this discovery.

Others at once began to investigate the high-altitude rays and there seemed no way of associating them with any specific heavenly body. They seemed to come from the cosmos generally, and so, in 1925, the American physicist Robert Andrews Millikan (1868–1935)‡ suggested that they be called "cosmic rays." It was a successful suggestion.

It was Millikan's notion that cosmic rays were electromagnetic in nature, that they were shorter still, and more energetic, than gamma rays were. He believed also that cosmic rays originated in the outskirts of the Universe where matter was being created. He considered cosmic rays a "birth cry" of matter, and said, "The Creator is still on the job." (Millikan, the son of a Congregationalist minister, was a sincerely religious man—as many scientists were and are.)

Not everyone agreed with Millikan. Some said that cosmic rays consisted of streams of very, very energetic particles, and, almost surely then, electrically charged particles, since even in the 1920s no penetrating particles had been discovered without electric charge.

Particles had won out over radiation in the case of cathode rays, while radiation had won out over particles in the case of X rays and gamma rays. How about cosmic rays?

The decision wasn't going to be easy. If cosmic rays were electromagnetic radiation, they would be *so* short in wavelength that even crystals wouldn't serve to produce diffraction effects. And if they were streams of electrically charged particles, they would be *so* energetic they would scarcely experience any deflection by any man-made magnetic field. Therefore, any experimental result was likely to be so borderline that it wouldn't settle the matter.

It occurred to some physicists, however, that cosmic rays, in reaching the Earth, had to pass through Earth's magnetic field. Earth's magnetic field was not terribly strong, but there were thousands and thousands of kilometers of it, and even a very small deflection should mount up and become noticeable.

If the cosmic rays were coming from every part of the sky equally, and if they consisted of charged particles, then Earth's magnetic field

‡He had received the Nobel Prize for physics in 1923 for his work in measuring the size of the electric charge on the electron.

ought to deflect them away from the magnetic equator (the region equidistant from the magnetic poles) and toward those poles. This is called the "latitude effect" since, in general, the effect of the Earth's magnetic field would be to shift the cosmic-ray incidence from the lower latitudes to the higher latitudes.

Attempts to demonstrate the latitude effect were not, at first, very convincing. Then, about 1930, the American physicist Arthur Holly Compton (1892–1962)* decided to go all out. He became a world traveler over a period of years, moving from place to place over the globe and measuring cosmic-ray intensity wherever he went.

In doing so, Compton was able to demonstrate conclusively that the latitude effect *did* exist, and that cosmic rays *were*, therefore, composed of electrically charged particles.

Millikan clung stubbornly to the electromagnetic version of cosmic rays, despite all the gathering evidence to the contrary, but he headed an ever-dwindling group. He proved wrong. No one doubts the particulate nature of cosmic rays today. It is known that they consist of positively charged particles, specifically of atomic nuclei, mostly hydrogen, but including nuclei up to those as heavy as iron, at least.

So the electromagnetic spectrum ends with the gamma rays at the short-wave end, as it ends with radio waves at the long-wave end. In the next chapter, then, we can finally turn to other subjects.

---

*He had received a share of the Nobel Prize for physics in 1927 for work he did on X rays.

CHEMISTRY

V

Big Brother

I was engaged in casual conversation with a young man the other day, and its natural course led him to remark that the south nave of the Cathedral of St. John the Divine was going to be constructed. (A nave, as I suppose you all know, is a long narrow hall, or corridor, that is part of a church.)

As soon as the young man said the word "nave," it occurred to me at once that if only the architect were named Hartz, and if a couple of ladies of the evening had managed to make it to the church before being caught by a pursuing policeman intent on an arrest, and if the ladies claimed sanctuary, then the policeman would be justified in declaiming:

> *"The nave of Hartz;*
> *It stole those tarts!"*

in parody of the well-known nursery rhyme concerning the "knave of hearts."

Since I am a devotee of wordplay, and since I admired this conceit I had invented, I felt I had to display it to the young man with whom I was conversing.

So I began, "Now if it should happen that the south nave were being constructed by an architect named Hartz—"

The young man said, "Yes, David Hartz."

"David Hartz?" I said, puzzled.

"Yes. Isn't he the one you're referring to? David Hartz of Yale, I think. Do you know him?"

"Are you serious? Is the architect really named Hartz?"

"*Yes!* You're the one who brought up his name."

What could I do? I was face to face with another Coincidence and it killed my piece of wordplay, which would have seemed drab by comparison. I didn't bother to trot it out.

Still, if the young man is correct about the name of the architect, then I trust that the structure will be known as "the nave of Hartz" through all eternity, and I will gladly bribe two young women of the appropriate profession to seek refuge there, so as to make my wordplay come true.

But coincidences are to be found not only in everyday life, but in science also, and thereby hangs a tale.

When chemists were studying the elements during the nineteenth century, they discovered a number of interesting similarities between one element and another. If no order existed among the elements, then those similarities would be merely unexplained (and, perhaps, unexplainable) coincidences, and they would make scientists as uncomfortable as fleas make a dog.

Chemists tried to find order and, in so doing, succeeded in establishing the periodic table of the elements (see "Bridging the Gaps," in *The Stars in Their Courses*, Doubleday, 1971).

Of all the elements in the list, carbon should be the dearest to us, since it is through its unusual (and, perhaps, unique) properties that life on Earth is possible.

In fact, we might even argue, if we were in a conservative mood, that carbon is the only conceivable basis of life *anywhere* in the Universe (see "The One and Only," in *The Tragedy of the Moon*, Doubleday, 1973).*

Yet how can carbon be unique? According to the periodic table, carbon does not stand alone, but is the head of a "carbon family"

---

*More radical views are possible, as in *Life Beyond Earth* by Gerald Feinberg and Robert Shapiro (Morrow, 1980), which I heartily recommend to all of you.

made up of chemically similar elements. The carbon family consists of five stable elements: carbon itself, silicon, germanium, tin, and lead.

Within a family, chemical similarities are strongest between elements adjacent to each other. This means that the element most similar to carbon in chemical properties is silicon, the next in line, and it is silicon with which this essay will deal.

Carbon has an atomic number of 6 and silicon one of 14. (By comparison, the atomic numbers of germanium, tin, and lead are 32, 50, and 82 respectively.) Silicon has an atomic weight of 28, compared to carbon's 12. The silicon atom is therefore $2\frac{1}{3}$ times as massive as is the carbon atom. Silicon is carbon's big brother, so to speak.

The atomic number tells us the number of electrons circling the nucleus of an intact atom. Carbon has six electrons divided into two shells: two electrons in the inner shell, four in the outer one. Silicon, on the other hand, has 14 electrons divided into three shells: two in the innermost, eight in the intermediate one, and four in the outermost. As you see, then, carbon and silicon each have four electrons in the shell farthest from the nucleus. We could describe carbon as (2/4) and silicon as (2/8/4) in terms of the electron content of their atoms.

When a carbon atom collides with another atom of any kind, it is the four electrons on the outskirts of the carbon atoms that interact with the electrons in the other atom in one way or another. It is this interaction that produces what we call a chemical change. When a silicon atom collides, it is again the four electrons on the outskirts that interact.

All electrons are identical down to the finest measurements scientists can make. The four outermost electrons of carbon and the four of silicon behave similarly, for that reason, and the chemical properties of the two elements are, therefore, also similar.

But in that case, if carbon, with four electrons on the outskirts of its atoms, has the kind of chemical properties that allow it to serve as the basis of life, ought not silicon, with *its* four electrons on the outskirts, also serve as a basis of life?

To answer that question, let's start at the beginning.

Silicon is an extremely common element. Next to oxygen, it is the most common component of Earth's crust. Some 46.6 percent of the total mass of the Earth's crust consists of oxygen atoms, and about

27.7 percent is silicon. (The other eighty elements that occur in the crust, taken altogether, make up the remaining 25.7 percent.) In other words, if we leave oxygen out of account, then there is more silicon in the Earth's crust than everything else put together.

Just the same, don't expect to stumble over a piece of silicon the next time you venture out into the world. It won't happen. Silicon is not to be found on Earth in its elemental form; that is, you won't find a chunk of matter made up of silicon atoms only. In the Earth's crust, all the silicon atoms that exist are combined with other kinds of atoms, chiefly those of oxygen, and therefore exist as "compounds."

For that matter, you can't pick up a hunk of Earth's crust and squeeze pure oxygen out of it, either, since the oxygen atoms present are combined with other kinds of atoms, chiefly silicon. There is considerable elemental oxygen in the Earth's atmosphere, but there is no naturally occurring free silicon to speak of anywhere in our reach.

Here we come across some differences between silicon and carbon. For one thing, carbon is not as common as silicon in the Earth's crust. For every 370 silicon atoms, there is only one carbon atom. (That still leaves carbon comparatively common, however.)

This is peculiar, since in the Universe as a whole, smaller atoms are more common than larger atoms (with some exceptions, for reasons that are understood) and carbon atoms are distinctly smaller than silicon atoms. In the Universe as a whole, astronomers estimate that there are seven carbon atoms for every two silicon atoms.

Why, then, is Earth's crust comparatively carbon-poor? —We'll let that go for now, but I promise I will return to this matter eventually.

Carbon, like silicon, is usually found in combination with other atoms, chiefly oxygen, but *unlike* silicon, sizable quantities of carbon are to be found in elemental form, as chunks of matter containing carbon atoms almost entirely. Coal, for instance, is anywhere from 85 to 95 percent carbon atoms.

But, then, coal originates from decaying plant material. It is the product of life. If carbon did not have properties that allowed it to serve as a basis for life, it would not occur in the free state in Earth's crust.

We might, conversely, argue that if silicon were enough like carbon to serve as the basis for another variety of life, it, too, would be likely to occur in the free state, when silicon life broke down. Consequently, if we find out why silicon won't serve as the basis of life, we will also find why it does not occur free as carbon does.

(As a matter of fact, the only reason oxygen occurs free in the atmosphere is because of the activity of plant life, which liberates oxygen as a side effect of photosynthesis. If life did not exist on Earth, the only elements that would occur free would be those that were particularly inert, chemically. Most of these, like helium or platinum, are very rare. The least rare of the inert elements is nitrogen and, as a result, there are sizable quantities of free nitrogen not only in the atmosphere of Earth, a planet rich in life, but also in the atmosphere of Venus, a totally barren planet!)

Even though nature has not been kind enough to prepare silicon in elemental form for us, chemists have learned to do it on their own. Two French chemists, Joseph Louis Gay-Lussac (1778–1850) and Louis Jacques Thénard (1777–1857) managed, in 1809, to decompose a silicon-containing compound and to obtain a reddish-brown material out of it. They did not examine it further. It is probable that this material was a mass of elemental silicon, though containing much in the way of impurities.

In 1824, the Swedish chemist Jöns Jakob Berzelius (1779–1848) obtained a similar mass of silicon by a somewhat different chemical route. Unlike Gay-Lussac and Thénard, however, Berzelius realized what he had, and went to considerable trouble to get rid of the nonsilicon impurities.

Berzelius was the first to get reasonably pure silicon, and to study what he had and report on its properties. For that reason, Berzelius is usually credited with the discovery of silicon.

Berzelius' silicon was "amorphous"; that is, the individual silicon atoms were arranged in an irregular fashion so that no visible crystals formed. (The word "amorphous" is from Greek, meaning "no shape," since crystals are distinguished by their regular geometrical form.)

In 1854, the French chemist Henri Étienne Sainte-Claire Deville (1818–81) prepared silicon crystals for the first time. These shone with a metallic luster, and this might make it seem that silicon is different from carbon in another important way, that silicon is a metal and carbon is not.

That, however, is not so. Although silicon has some properties that are similar to those of metals generally, it has others that are not, and it is therefore a "semimetal." Carbon, in the form of graphite, also has some metallic properties (it conducts electricity moderately well, for instance). The two elements, consequently, are not startlingly different in this respect.

Carbon atoms, to be sure, are not bound to arrange themselves in the manner that produces graphite. They may also arrange themselves in a more compact and symmetrical manner in order to produce diamond, which shows no metallic properties whatever (see "The Unlikely Twins," in *The Tragedy of the Moon*, Doubleday, 1973).

Diamond is particularly notable for being hard and, in 1891, the American inventor Edward Goodrich Acheson (1856–1931) discovered that carbon, when heated with clay, yields another very hard substance. Acheson thought the substance was carbon combined with alundum (a compound of aluminum and oxygen atoms, both of which are found in clay).

He therefore called the new hard substance "carborundum."

Actually, carborundum turned out to be a compound of carbon and silicon (silicon atoms are also found in clay). The compound consisted of silicon and carbon atoms in equal quantities ("silicon carbide," or, in chemical symbols, SiC). This mixture of atoms took on the compact and symmetrical arrangement that occurs in diamond.

In carborundum, the carbon and silicon atoms are placed alternately within the crystal structure. The fact that one can substitute silicon atoms for every other carbon atom and still have a very hard substance shows how similar the two elements are. (Not all properties are preserved, however. Carborundum does not have the transparency or beauty of diamond.)

Carborundum, incidentally, is not quite as hard as diamond. Why is that?

Well, silicon and carbon atoms are chemically similar, thanks to the fact that both have four electrons in the outermost shell, but they are not identical. The silicon atom has three electron shells compared to the carbon atom's two. That means the distance from the outermost shell of the silicon atom to its nucleus is greater than in the case of the carbon atom.

The electrons carry a negative electric charge and are held in place by the attraction of the positive charge on the atom's nucleus. This attractive force decreases with distance and is therefore weaker in the silicon atom than in the smaller carbon atom.

Furthermore, between the outermost electrons and the nucleus in the silicon atom are the ten electrons of the two inner shells, but in the case of carbon, there intervene only the two electrons of the lone inner shell. Each negatively charged inner electron, existing between the outermost shell and the nucleus, tends to neutralize the nucleus' positive charge somewhat and weakens the nucleus' hold on the outermost electrons.

When two carbon atoms cling together, that is because of the attractive force generated by the association of two electrons (one from each atom). The more firmly those two electrons are held by the respective nuclei of the two atoms, the stronger the bond between them.

Therefore, the carbon-carbon bond is stronger than the silicon-silicon bond, and the silicon-carbon bond should be of intermediate strength.

One way of demonstrating this is by melting point. As the temperature rises, the atoms vibrate more and more strongly until, finally, they break the bonds holding them together and slide over each other freely. The solid has become a liquid. The tighter the bonds, therefore, the higher the melting point must be.

Carbon doesn't actually melt, but "sublimes"; that is, it turns from a solid into a vapor directly, but we'll call that the melting point just the same. The melting point of carbon is over 3500° C while that of silicon is only 1410° C. Carborundum (which, like carbon, sublimes) has the intermediate melting point of 2700° C.

Again, you can judge the tightness of the bond by the hardness of the substance. The stronger the bond between the atoms, the more the substance resists deformation, and the more easily it inflicts deformation (in the form of scratches, for instance) upon other, softer substances.

Diamond is the hardest substance known. Carborundum is not quite as hard, but is harder than silicon.

Despite the fact that carborundum is not quite as hard as diamond, it is more useful as an "abrasive" (something which is hard enough to wear down, through friction, softer objects, without itself being much affected). Why?

The answer is a matter of price. We all know how rare and expensive diamonds are, even impure ones of less-than-gem quality. Carborundum, on the other hand, can be made out of ordinary carbon and clay, both of which are about as cheap as anything can reasonably be expected to be.

I said, earlier, that silicon atoms are, in nature, most frequently found in combination with oxygen atoms. The oxygen atom is readily able to accept two electrons from another atom, combining each electron it accepts with one of its own. Two electron pairs are formed between the two atoms and this is called a "double bond," which we can represent in the following fashion: "Si=O." The silicon atom has four outer electrons, however, and it is perfectly capable of donating two electrons to each of two different oxygen atoms.

The result is O=Si=O, which can also be represented, more simply, as SiO_2, and which can be called "silicon dioxide." It is an old-fashioned habit, arising in the days when chemists did not know exactly how many atoms of each element were present in combination (or that there were atoms at all), to have the name of a compound of some element with oxygen end with a final "a." Consequently, silicon dioxide is also called "silica."

In fact, silica was the name that was used first, and the final "a" indicated that it was suspected of being a combination of oxygen with an element that had not yet been isolated. Once the other element was obtained, it was named "silicon" from silica, the "n" ending being conventional for a nonmetallic element, as in boron, hydrogen, and chlorine.

The purest form of silica, when it contains virtually nothing but silicon and oxygen atoms, is best known as "quartz," a word of unknown origin.

The astonishing thing about quartz, if it is pure enough, is that it is transparent. There are very few naturally occurring solids that allow light to pass through with scarcely any absorption, and quartz is one of those few.

The first such substance that early man encountered was ice, which, if it is formed slowly, and in a reasonably thin layer, is transparent. When men who had encountered ice later encountered quartz, they could only think they had found another form of ice, one which had formed in so superrigid a manner under such supercold conditions that it was no longer capable of melting.

The Greeks, therefore, called quartz *krystallos*, which was their word for "ice." This became *crystallum* in Latin and "crystal" in English. The prefix "cry" is still used to mean "very cold," as in "cryogenics" (the production of ultralow temperatures); "cryonics" (the preservation of living tissue by ultralow temperatures), "cryometer" (a thermometer for registering ultralow temperatures), and so on.

"Crystal," as an English word for "ice," is, however, obsolete now. It is more often used to signify a transparent object, even when it is not made of quartz. For instance, we still talk of a fortune teller looking into her "crystal ball," which is, of course, simply glass.

Then, too, when the Greeks said that each planet was part of a sphere, and turned with its sphere, those came to be spoken of as "crystalline spheres" because of their transparency. (They were *totally* transparent, for they didn't exist.)

Quartz was usually found in straight-line, smooth-plane, sharp-angled shapes, and the word "crystal" came to mean that. Naturally occurring solids of such shapes came to be called "crystals," whether they were quartz or not.

Quartz is not necessarily transparent, because it is not necessarily sufficiently pure. If the impurity is not very great, the quartz may stay transparent, but gain a color. The best and most beautiful example of that is the purple "amethyst."

(The ancient Greeks, noting amethyst's wine color, reasoned, by the principles of sympathetic magic, that it must counter the effects of wine. Wine drunk from an amethyst cup, they were sure, would taste great but would not intoxicate. In fact, "amethyst" is from Greek words meaning "no wine." Don't bother to try it; it won't work.)

With greater amounts of impurity, you have silica that is chemically combined with such metals as iron, aluminum, calcium, potassium, and so on—or mixtures of several of these. Such compounds are referred to as "silicates" and are, for the most part, dull and opaque substances. Included among the silicates are granite, basalt, clay, and so on. Indeed, Earth's rocky crust, together with the mantle beneath, is largely silicate in nature.

Flint is a common silicate, and it was very important to early man, because it could be chipped or ground into sharp edges and points and was therefore the best thing for tools like knives, hatchets, spear points, and arrow points in any society that lacked metals. The word "flint" is from an old Teutonic term meaning a "rock chip," which was what you got when you worked flint into a tool. The rock chip itself was sometimes the tool.

The Latin word for "flint" is *silex*, and if one wished to speak of something that was made "of flint," the genitive form of the word, *silicis*, was used. It was from flint, then, that we got first "silica" for silicon dioxide, and then "silicon" for the element.

Shattered bits of quartz, shattered, usually, by the action of waves on a shore, form "sand." The color of sand depends upon the purity of the quartz and, if not pure, on the nature of the impurities. Pure quartz will produce a rather white sand; the usually sandy color of sand (what else?) is due to iron content.

The oxygen atom is smaller than the silicon atom, but larger than the carbon atom. Therefore silicon dioxide ought to have a higher melting point than silicon itself does, but a lower one than carborundum.

That, indeed, is the way it works out. Silicon dioxide has a melting point of about 1700° C, which is higher than that of silicon and lower than that of carborundum.

If appropriate substances that contain sodium and calcium atoms are added to sand, and if the mixture is heated, it melts and becomes "glass," which is, essentially, a sodium-calcium silicate. Other substances can, however, be added to gain certain desired qualities such as color, or hardness, or resistance to temperature change, or limpid transparency.

Glass is, on the whole, as transparent as quartz to visible light, but glass is much more useful in most practical ways.

For one thing, glass can be made out of sand, which is much more common than intact crystals of quartz, and so is much cheaper than quartz. For another, glass melts at a lower temperature than quartz, so that it is easier to work with.

Then, too, glass does not really solidify, but remains a liquid. That liquid, however, gets stiffer and stiffer as it cools, until it is a solid to all intents and purposes. The glass we routinely handle is, in short, a liquid because its atomic arrangement is random as in liquids, rather than orderly as in solids; yet it has the rigidity of a solid. This means that glass has no sharp melting point but remains a sort of gooey liquid over a fairly large temperature range, and this, again, makes it easier to work with.

Now, then, can carbon substitute for silicon and produce carbon analogs of quartz, sand, and rock, of silica and silicates?

Carbon can make a good beginning. It, too, can donate two of its four outer electrons to each of two oxygen atoms. The result is O=C=O, or CO_2, which is universally known as "carbon dioxide" and which, from the formula, certainly seems to be an analog of silicon dioxide.

The bond between carbon and oxygen atoms, all things being equal, is stronger than the bond between silicon and oxygen, since carbon atoms are smaller than silicon atoms. Therefore it is only fair to suppose that carbon dioxide will melt at a higher temperature than silicon dioxide will.

Carbon dioxide has a melting point of $-78.5°$ C (though, in actual fact, it sublimes rather than melts), and this is 1800 degrees *lower* than the melting point of silicon dioxide.

Why is that? —Well, the answer to that, and to the matter of carbon's comparatively low occurrence on Earth, and to the big question of which will lead to life, and why, must await the next chapter.

VI
Bread and Stone

In the Sermon on the Mount, Jesus assured his listeners that God the Father would be kind to humanity. He demonstrated that by pointing out that human fathers, vastly imperfect by comparison, were kind to *their* children. He said:

". . . what man is there of you, whom if his son ask bread, will he give him a stone?" (Matthew 7:9).

A bitter echo of this verse was heard eighteen centuries later in connection with Robert Burns, the great Scottish poet, who lived and died in grinding poverty, even while he was turning out his now world-famous lyrics.

After his death in 1796, at the age of thirty-seven, the Scots discovered that he was a great poet (it is always easier to honor someone after it is no longer necessary to support him) and decided to erect a monument to him. This was told to Burns's aging mother, who received the news with less than overwhelming gratitude.

"Rabbie, Rabbie," she was reported to have said, "ye asked for bread, and they gave ye a stone."

I love the story and it brings a tear to my eye every time I tell it, but like all the historical stories I love, it may be apocryphal.

The English satirist Samuel Butler, best known for his poem "Hudibras," died in dire want in 1680, and, in 1721, Samuel Wesley, after noting Butler's monument in Westminster Abbey, wrote:

> *A poet's fate is here in emblem shown:*
> *He asked for Bread, and he received a Stone.*

It's unlikely that Mrs. Burns, three quarters of a century later, was quoting Wesley, but it seems to me likely that whoever reported Mrs. Burns's remark was really quoting.

In any case, bread is the product of the carbon atom, and stone is the product of the silicon atom. And though carbon and silicon are very similar in atomic structure, their products are so different that they form a natural and powerful antithesis.

I ended the previous chapter by comparing carbon dioxide and silicon dioxide, the former turning into a gas at so low a temperature that it remains a gas even in the extremest wintry weather of Antarctica, the other turning into a gas at so high a temperature that even the hottest volcanoes don't produce any significant quantities of silicon dioxide vapors.

In the molecule of carbon dioxide, each carbon atom is combined with two oxygen atoms, thus $O=C=O$. The carbon atom (C) is attached to each oxygen atom (O) by a double bond, that is, by two pairs of electrons. Each atom participating in such a double bond contributes one electron to each pair, or two electrons altogether. The oxygen atom has only two electrons to contribute under normal circumstances, the carbon atom, four. The carbon atom, therefore, forms a double bond with each of two oxygen atoms, as shown in the formula.

The silicon atom (Si) is very similar to the carbon atom in its electron arrangements, and it, too, has four electrons to contribute to bond formation. It, too, can form a double bond with each of two oxygen atoms, and silicon dioxide can be represented as $O=Si=O$.

In the previous chapter, I pointed out that the bonds holding carbon and oxygen together are stronger than those holding silicon and oxygen together, and suggested this meant that carbon dioxide should have higher melting and boiling points than silicon dioxide has. The reverse is, in actual fact, true, and I posed this as a problem.

Actually, I was oversimplifying. There are indeed times when melting points and boiling points signify the breaking of strong bonds between atoms, so that the stronger the bonds the higher the melting and boiling points. This is true when each atom in a solid is held to its neighbors by strong bonds. There is then no way of converting the solid to, first, a liquid and then a gas, but by breaking some or all of those bonds.

In other cases, however, two to a dozen atoms are held together strongly to form a discrete molecule of moderate size, and the individual molecules are bound to each other weakly. In that case, melting and boiling points are reached when those weak intermolecular bonds are broken, and the individual molecules are freed. In that case, the strong bonds within the molecule need not be touched, and the melting points and boiling points are then usually quite low.

In the case of a boiling point, in particular, we have a situation in which the intermolecular bonds are completely broken so that a gas is produced in which the individual molecules move about freely and independently. At a sublimation point, the intermolecular bonds in a solid are completely broken to form a gas which would be made up of completely independent molecules.

The boiling point of silicon dioxide is about 2300° C, while the sublimation point of carbon dioxide is −78.5° C. Clearly, in heating silicon dioxide to a gas, we must break strong bonds between atoms; while in heating carbon dioxide to a gas, we need break only weak intermolecular bonds.

Why? The formulas, $O=Si=O$ and $O=C=O$, look so similar.

To begin with, we must understand that a double bond is *weaker* than a single bond. This seems to go against common sense. Surely, a grip with both hands would be stronger than a grip with only one hand. Holding something with two rubber bands, two ropes, two chains would seem a stronger situation than with only one in each case.

Nevertheless, this is not so in the case of interatomic bonds. To explain that properly requires quantum mechanics, but I will do everyone a favor* by offering a more metaphoric explanation. We can imagine that there's only so much space between atoms, and that when

* Chiefly myself!

four electrons crowd into the space to set up a double bond, they don't have enough room to establish a good grip. Two electrons, setting up a single bond, do better. Imagine yourself squeezing both hands into a constricted space, and holding on to something with thumb and forefinger of each hand. Inserting one hand and achieving a good all-fingers grip would be far more effective.

Consequently, if there is a chance to rearrange the electrons in silicon dioxide in such a way as to replace the double bonds by single bonds, the tendency will be for that to happen.

Where there are many silicon dioxide molecules present, for instance, each oxygen atom distributes its electrons so that it holds on to each of two different silicon atoms with a single bond apiece, rather than to one silicon atom with a double bond. Instead of $O=Si=O$, you have $-O-Si-O-Si-O-Si-$ and so on, indefinitely, in either direction.

Each silicon atom has four electrons to contribute and can form four single bonds, but each uses only two single bonds in the chain just pictured. Each silicon atom, therefore, can start an indefinite chain in two other directions, so that you end up with—

$$
\begin{array}{ccccccc}
 & | & & | & & | & \\
-\text{O}- & \text{Si} & -\text{O}- & \text{Si} & -\text{O}- & \text{Si} & -\text{O}- \\
 & | & & | & & | & \\
 & \text{O} & & \text{O} & & \text{O} & \\
 & | & & | & & | & \\
-\text{O}- & \text{Si} & -\text{O}- & \text{Si} & -\text{O}- & \text{Si} & -\text{O}- \\
 & | & & | & & | & \\
 & \text{O} & & \text{O} & & \text{O} & \\
 & | & & | & & | & \\
-\text{O}- & \text{Si} & -\text{O}- & \text{Si} & -\text{O}- & \text{Si} & -\text{O}- \\
 & | & & | & & | & \\
\end{array}
$$

This looks two-dimensional, but it isn't really. The four bonds of silicon are distributed toward the four apices of a tetrahedron and the result is a three-dimensional structure, rather like that in diamond or silicon carbide.

Consequently, each chunk of pure silicon dioxide ("quartz") is, in effect, an enormous molecule, in which there are, on the whole, two oxygen atoms for every silicon atom. To melt and boil such a chunk requires the breaking of the strong Si—O bonds, so that we have a high boiling point and don't encounter gaseous silicon dioxide under Earth-surface conditions.

All this remains true if other types of atoms join the silicon-oxygen lattice in numbers that are not large enough to disrupt that lattice completely, thus forming silicates. These silicates, generally, are high-melting and high-boiling.

The matter is quite different with carbon dioxide. Smaller atoms tend to form stronger bonds, so that the carbon atom, which is smaller than the silicon atom, bonds more strongly with oxygen than silicon does. In fact, even the C=O double bond, while weaker than the C—O single bond, is nevertheless sufficiently strong so that the tendency to distribute itself into single bonds is much smaller than in the case of silicon dioxide. There are certain advantages in stability of small molecules over large, and this, combined with the comparative strength of the carbon / oxygen double bond, tends to keep carbon dioxide in the form of small molecules.

If the temperature is low enough, the individual molecules of carbon dioxide cling together and form a solid, but the molecules are held together by relatively weak intermolecular bonds and these are easily broken. Hence, the low sublimation point.

Other atoms can combine with carbon dioxide to form "carbonates," and these remain solid at Earth-surface temperatures. If heated to higher temperatures, however, they break up and give off carbon dioxide gas at considerably lower temperatures than the boiling point of silicates.

Calcium carbonate ("limestone") will, for instance, give off carbon dioxide gas at about 825° C.

When a planetary system forms, the process of formation produces a hot planet to begin with. If the forming planets are comparatively near the central sun, the temperature rises still higher as a result.

Under those conditions, the only solids one can have are those which consist of atoms forming large atom lattices and which, therefore, have high melting and boiling points. This includes two varieties of substances, which tend to separate as the planet develops: metals (chiefly iron, plus those metals that will mix with it relatively freely) and silicates.

The dense metals tend to collect at the center of the planet, with the lighter silicates surrounding that core as an outer shell.

This is the general structure of the five worlds of the inner Solar System: Mercury, Venus, Earth, Moon, and Mars. (In the case of Mars and the Moon, the metallic component is quite low.)

Those elements whose atoms fit with difficulty, or not at all, into the metallic or silicate lattice, tend to be left over as individual atoms, as small molecules, or as lattices in which the atoms are but feebly held together. In all cases, they are low-melting (''volatile'') and, in the early days of planetary formation, existed to a large extent as vapors.

Since the metals and silicates are made up of elements that, in turn, make up a relatively small fraction of the original materials out of which planetary systems form, the worlds of the inner Solar System are comparatively small, and possess correspondingly weak gravitational fields—too weak to hold vapors.

This means the loss of most or all of some of the elements that are particularly common in the original preplanetary mixture: hydrogen, helium, carbon, nitrogen, neon, sodium, potassium, and argon.

Thus, Mercury and the Moon possess little or no hydrogen, carbon, and nitrogen, three elements without which life as we know it cannot exist. Venus and Earth are massive enough to have hung on to some of these elements, and each has a substantial atmosphere of volatiles. Mars, with a weaker gravitational field (it has only one tenth the mass of Earth) was, because of its greater distance from the Sun, cool enough to hang on to a tiny quantity of volatile matter, and has a thin atmosphere.

Beyond Mars, in the outer Solar System, the planets remained cool enough to collect substantial proportions of those volatiles which made up 99 percent of the original mixture (chiefly hydrogen and helium), so that they grew large and massive. As they grew, their gravitational field intensified, and they were able to grow still more rapidly (the ''snowball effect''). The result were the large outer planets: Jupiter, Saturn, Uranus, and Neptune. These are the so-called gas giants, made up chiefly of a mixture of hydrogen and helium, with small molecules containing carbon, nitrogen, and oxygen as substantial impurities, and with (it is presumed) relatively small cores of silicates and metals at the center.

Even the smaller worlds of the outer Solar System became cool enough, at an early stage, to collect volatiles. The molecules of these contain carbon or nitrogen or oxygen, each in combination with hydrogen. At the present very low temperatures of these worlds, these volatiles are in the solid state. They are "ices," so called because of their general resemblance in properties to the best-known example here on Earth—frozen water.

The four large satellites of Jupiter, as an example, have undergone heating through Jupiter's tidal effect (which grows rapidly greater as the distance of the satellite is smaller). Ganymede and Callisto, the two farthest satellites, have undergone little heating and are essentially icy worlds, and larger than the other two. Io, the innermost, has been too warm to collect volatiles and is essentially silicate in nature, while Europa, which lies between Io and Ganymede, seems to be silicate, with an icy cover.

But let's get back to Earth, which consists of a liquid nickel-iron core surrounded by a silicate mantle.

On the very surface are those volatiles that Earth has managed to keep. Hydrogen atoms are to be found chiefly as part of the water molecules that make up our (comparatively) vast oceans. Nitrogen atoms are to be found as two-atom molecules in the atmosphere. Carbon atoms are found as carbon dioxide in the atmosphere (in small quantities), as carbonates in the crust, and as elementary carbon in the form of coal deposits.

Earth is, however, depleted in these elements and, while they are present in sufficient quantity to allow a complete and diverse load of life, there are less of these elements on Earth by far than there is in an equal mass of matter more representative of the overall composition of the Universe (say, in an equal mass of Jupiter or the Sun).

But if the Earth's crust contains 370 silicon atoms for every carbon atom, and the two are so similar in many of their chemical properties, why should life form about the carbon atom rather than the silicon atom?

In this connection, we have to remember that life is a rather complex atomic dance. Life represents a relatively low entropy system, maintained against an overwhelming tendency ("the second law of thermodynamics") to raise the entropy. Life is made up of very complex and fragile molecules that, left to themselves, would break down.

It contains high concentrations of certain types of atoms or molecules in some places, low concentrations in others; where, left to themselves, the concentrations would promptly begin to even out—and so on.

In order to maintain the state of low entropy, the chemistry of life keeps up an unceasing activity. It is not that molecules don't break down or uneven concentrations don't even out; it is that the complex molecules are built up again as fast as they break down, and the concentrations made uneven as fast as they even out. It is as though we were keeping a house dry during a flood, not by stopping the flood (which we can't do) but by assiduously and tirelessly sweeping out the water as fast as it pours in.

This means that there must be a constant shuffling of atoms and molecules, and that the basic raw materials of life must exist in a form that enables them to be seized and used rapidly. The raw materials must exist as small molecules in considerable quantity under conditions that enable the bonds holding the atoms together within the molecules to be easily broken and reformed, so that molecules of one type are forever being converted to another.

This is accomplished by the use of a fluid medium in which the various molecules are dissolved. There they are present in high concentration, they move about freely, and they serve the purpose. The fluid medium used is water, which is plentiful on Earth, and which is a good solvent for a wide variety of substances. Life as we know it, in fact, is impossible without water.

The molecules that are useful for life are those that are soluble in water, therefore, or that can be made soluble. Carbon dioxide, for instance, is fairly soluble in water. Oxygen is only slightly soluble but it attaches readily to hemoglobin so that the small quantity that does dissolve is snatched up at once, leaving room for another small quantity to dissolve, and so on.

The process of solution in water is, however, similar in some ways to the processes of melting and boiling. Interatomic or intermolecular bonds must be broken. If you have a whole lattice of atoms, the entire lattice will not enter solution as an intact mass. If the lattice can be pulled apart into smaller fragments, on the other hand, those fragments may be dissolved.

Silicates form a tightly bound lattice, and the bonds are as resistant to breakage by water as by heat. Silicates are "insoluble"—which is a good thing, or the oceans would dissolve much of the continental areas and produce a thick sludge, which would be neither sea nor land and in which life, as we know it, could not exist.

But this also means that silicon atoms do not exist in the form of small, soluble molecules and, consequently, are not incorporated into actively living tissue. Silicon, therefore, does not serve as the basis of life, and carbon does.

That, however, is under Earthly conditions. What about other conditions?

A planet's chemical condition can be "oxidizing" or "reducing." In the former case, there is a preponderance of atoms that accept electrons, as would be the case with large quantities of free oxygen in the atmosphere. In the latter case, there is a preponderance of atoms that give up electrons, as would be the case with large quantities of free hydrogen present in the atmosphere. Earth has an oxidizing atmosphere; Jupiter, a reducing one. Originally, Earth may have had a reducing atmosphere, too.

In an oxidizing atmosphere, carbon tends to exist as carbon dioxide. In a reducing atmosphere, it tends to exist as "methane," the molecule of which consists of a carbon atom to which four hydrogens are attached (CH_4). In the outer Solar System, where reducing conditions are the rule, methane is as extraordinarily common.

Methane is the parent of an endless number of other substances, for carbon atoms can easily attach to each other in chains or rings, with any spare bonds connected to hydrogen atoms. There are thus an enormous number of possible "hydrocarbons," with molecules of various size made up of carbon and hydrogen only. Methane is merely the simplest of these.

Add an occasional atom of oxygen, nitrogen, sulfur, or phosphorus (or combinations of these) to the basic hydrocarbon skeleton, and you have the vast number and variety of compounds found in living organisms ("organic compounds"). These are all, after a fashion, elaborations of methane.

In short, the chemicals of life are of the type one would expect to be formed under reducing conditions, and that is one of the reasons that chemists suspect that the early Earth, at the time life came into existence, had a reducing atmosphere or, at the very least, not an oxidizing one.

Silicates are, however, characteristic of an oxidizing environment. Might not silicon form other kinds of compounds under reducing conditions? Could not silicon, like carbon, combine with four hydrogen atoms?

The answer is: Yes. The compound, SiH_4, does exist and is called "silane."

Methane has a boiling point of $-161.5°$ C, so under Earth-surface conditions, it is always a gas. Silane is quite similar in properties, with a boiling point of $-112°$ C, so that it, too, is a gas. (The boiling point of silane is distinctly higher than that of methane, because its molecular weight is distinctly higher, 28 as compared to 16.)

Then, just as carbon can form chains, with hydrogen taking up the spare bonds, so can silicon.

A two-carbon chain can add on six hydrogen atoms; a three-carbon chain, eight hydrogen atoms; and a four-carbon chain, ten hydrogen atoms. In other words, we can have C_2H_6, C_3H_8, and C_4H_{10}, which are called "ethane," "propane," and "butane" respectively. (Each name has a rationale behind it, but that's another story for another day.)

Similarly, we have Si_2H_6, Si_3H_8, and Si_4H_{10}, which are called "disilane," "trisilane," and "tetrasilane" respectively.

The carbon compounds have boiling points of $-88.6°$ C, $-44.5°$ C, and $-0.5°$ C respectively, so that all three are gases under Earth-surface conditions, though butane would be a liquid under ordinary winter conditions, and propane would be a liquid under polar conditions.

The silanes have appropriately higher boiling points. Disilane has a boiling point of $-14.5°$ C, trisilane, one of 53° C, and tetrasilane, one of 109° C. Under Earth-surface conditions, disilane is a gas, while trisilane and tetrasilane are liquids.

All this looks very hopeful, but there must be a catch, and there is. A single bond between carbon and oxygen has an energy content of 70 kilocalories per mole (and we can take the unit for granted henceforward), while the energy content of the bond between carbon and hydrogen is 87. There is a tendency for carbon to remain bonded to hydrogen, therefore, even in the presence of lots of oxygen. The hydrocarbons are quite stable under Earth-surface conditions.

Gasoline and paraffin are both mixtures of hydrocarbons. The former can burn in an automobile engine and the latter in a candle, but the burning has to be initiated. Left to itself, gasoline and paraffin will remain as such for extended periods of time.

The silicon-oxygen bond is, however, 89 and the silicon-hydrogen bond is 75. That means that there is a tendency for silicates to remain silicates even under reducing conditions, whereas silanes are comparatively easily oxidized to silicates.

In short, the odds are weighted in favor of hydrocarbons in the case of carbon, and in favor of silicates in the case of silicon. Given the slightest excuse, carbon will be converted to hydrocarbons and life, while silicon will be converted to silicates and nonlife.

In fact, even if silanes *were* formed, the result would probably not be life. Life requires very complicated molecules, and carbon atoms can combine into very long chains and very complicated sets of rings. That is because the carbon-carbon bond is quite strong—58.6. The silicon-silicon bond is distinctly weaker—42.5.

This means that a chain of silicon atoms is feebler than one of carbon atoms and falls apart more readily. In fact, chemists have not been able to form anything more complicated than a hexasilane, with six silicon atoms in the molecule. Compare this with the carbon chains in ordinary fats and oils, which are commonly made up of 16 carbon atoms linked together—and that is by no means a record.

Furthermore, carbon atoms cling together strongly enough to make possible the existence of carbon-carbon double bonds, and even triple bonds, though these are weaker than single bonds. This multiplies and remultiplies the number and variety of organic compounds that are possible.

Double and triple bonds have been thought to be not possible in the case of silicon-silicon combinations, so that whole masses of complexity were removed from potential existence.

But only *apparently*. In 1981, double bonds involving the silicon atom were, for the first time, reported. These were not in silanes, but in other types of silicon compounds that (perhaps?) might serve as the basis of life.

For a discussion of that, however, see the next chapter.

VII

A Difference of an "E"

When someone writes as much as I do, he must live in constant dread of possibly seeming, now and then, to have borrowed the work of others in an unauthorized manner. I say "seeming" because there is, of course, no danger in my case of *actually* doing so. Seeming, however, is bad enough.

For instance, some years ago an encyclopedia asked me to go over their short article on science fiction in order to correct and update it. I read it and it seemed perfectly correct, so I made no changes. I did add two sentences as updating, collected my small fee, and forgot the whole thing.

What I did not know was that the encyclopedia then removed the name of the original author and placed my own name on the article instead. About a month ago, I received a hot letter from a reader who was convinced he had uncovered some skulduggery. He sent me photocopies of the earlier article over the original author's name, and of the later article over my name, and, in a most offensive manner, demanded an explanation.

Patiently, I explained and said I had not known of the auctorial substitution. (After all, my record of authorship is not so small that I have to appropriate a five-inch encyclopedia item.)

—It didn't help. Still breathing flame, my investigating correspondent wrote to the encyclopedia for *their* version of the story.

Here's an example with more substance to it. About twenty years ago, at an MIT picnic, I heard a humorous retort to the well-known, sloppy-sentimental song that goes "Tell me why the stars do shine / Tell me why does the ivy twine / Tell me why the skies are blue / And I will tell you why I love you."

The MIT version of the second verse was:

> *Nuclear fusion makes stars to shine.*
> *Tropisms make the ivy twine.*
> *Rayleigh scattering makes skies so blue.*
> *Glandular hormones is why I love you.*

I liked the verse and, given my memory, remembered it, and sang it, year in and year out. Finally, I included it (slightly modified) in my book *Isaac Asimov's Treasury of Humor*. I laid no claim to it, you understand, but neither did I give credit to anyone, for I did not know who had written it.

A couple of months ago, I received a letter from Richard C. Levine, now living in Texas, who picked up the *Treasury* and found his own quatrain staring at him. I gladly now give credit.

It works the other way around, too. A physicist, last week, told me that the most exciting cosmogonic theory now being worked on was the creation of the Universe from *nothing* and told me the basic reasoning behind it. I asked when that had first been suggested. Back about 1972 or 1973 by so-and-so, said he.

With great satisfaction, I referred him to my essay "I'm Looking Over a Four-Leaf Clover" (see *Science, Numbers, and I*, Doubleday, 1968), which advanced a similar theory based on similar reasoning, and later sent him a copy.

Of course, I assumed nothing out of the way. I am sure that the physicist of 1972 arrived at his conclusions independently, in far greater detail, and with far greater precision of argument than I was capable of doing—but I was delighted to be able to claim a certain amount of priority, that's all.

I was also pleased that something that I had presented as nothing more than a delightful bit of speculation was coming to be of serious concern to physicists. That is not an unusual thing in the history of science, actually—which brings me to the English chemist Frederick Stanley Kipping (1863–1949).

Kipping was interested in asymmetric molecules, something I discussed in some detail in "The 3-D Molecule" (see *The Left Hand of the Electron*, Doubleday, 1972). To repeat the point briefly here, if a carbon atom is attached to four different atoms or atom groupings, the resulting molecule can be arranged in one of two different ways, one the mirror image of the other. Such molecules are asymmetric.

The nature and reason for the asymmetry was explained in 1874, and there was no reason why the explanation should apply to the carbon atom only. Kipping, together with his assistant, William Jackson Pope (1870–1939), labored to synthesize asymmetric molecules involving such atoms as nitrogen and tin.

In 1899, Kipping began a long series of researches on silicon compounds, supposing, quite rightly, that the silicon atom, which is so similar chemically to the carbon atom, should produce asymmetric molecules under the same conditions that the carbon atom did.

This brings us back to the subject matter of the preceding two essays.

The silicon compounds occurring in nature are the silicates, which, as I explained in the previous chapter, have silicon atoms connected to oxygen atoms by each of their four bonds. Under these conditions, asymmetry is not to be expected at the molecular level.

What Kipping wanted to do was to connect silicon atoms to different groupings at different bonds.

As it happened, a French chemist, François Auguste Grignard (1871–1935), had, in 1900, worked out a way of attaching atom groups to other atoms by means of the use of magnesium metal in dry ether.

Making use of such "Grignard reactions," Kipping began to add atom groupings to silicon atoms in unprecedented ways, and tried to synthesize molecules in which silicon atoms replaced key carbon atoms in simple and well-known carbon compounds.

Consider carbon dioxide, for instance, $O=C=O$. Leave one oxygen in place, but remove the other and substitute for it two different carbon-containing atom groups, one for each of the carbon bonds. Symbolize the two carbon-containing groups as R_1 and R_2. You have a molecule that looks like this:

$$R_1 \diagdown$$
$$\qquad C = O.$$
$$R_2 \diagup$$

Chemists call this a "ketone," and the names given to such compounds usually have an "=one" suffix to indicate that.

Kipping tried to synthesize the silicon analog of a ketone,

$$R_1 \diagdown$$
$$\qquad Si = O,$$
$$R_2 \diagup$$

and, using the usual suffix, referred to such an analog as a "silicone."

Personally, I don't like the name and wish he hadn't thought of it. "Silicone" and "silicon" are distinguished by a difference of an "e," and that isn't enough. It's too easy to produce confusion by means of a typographical error.

What's more, Kipping never managed to produce a silicon-ketone. Yet the name, "silicone," came to be applied not only to that molecule, but to all compounds in which silicon atoms are attached to carbon-containing atom groups.

Kipping worked on the problem of asymmetric silicon molecules for forty years and published fifty-one papers. He achieved his aims, and was able to show, quite thoroughly, that silicon compounds followed the same rules of asymmetry that carbon compounds did.

The silicones he formed in demonstrating this did not seem good for anything else, however. They were curiosities that served only to make a theoretical point. As late as 1937, Kipping, then seventy-four, said, sadly, "The prospect of any immediate and important advance in this section of organic chemistry does not seem to be very hopeful."

He was wrong! In 1941, the first silicone patents were issued, and a silicone industry began to grow and expand very rapidly. Kipping lived to be eighty-five and had the pleasure of witnessing this unexpected success. His ivory-tower research had become a matter of practical value after all.

Let's see why.

The silicon dioxide lattice, as I showed you in the previous chapter, looks something like this (if you remember that it is actually three-dimensional and not two-):

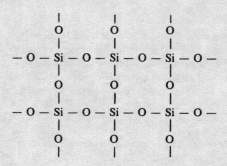

These attachments can go on indefinitely, so that silicon dioxide (as well as silicates, which have metal atoms added to the lattice here and there) is a high-melting solid.

Suppose, though, that in place of the oxygen atoms that serve as the connecting bridges between the silicon-oxygen chains, you have some atom, or group of atoms, that has but one bond. It can attach itself to the silicon atom of a particular chain with its one bond and will then have nothing left over to attach to a neighboring chain.

Let's consider carbon-containing atom groups in this connection. Carbon atoms (like silicon atoms) possess four bonds, but three of them can be firmly attached to hydrogen atoms, leaving the fourth bond free for attachment elsewhere. We might represent the group thus: $-CH_3$. It is called a "methyl group." (I have the urge to explain here why it should be called "methyl," but I will suppress it for now and succumb to the temptation some other time in another essay.)

If you imagine a methyl group replacing an oxygen atom in the silicate lattice, it would break a bridge between two silicon atoms:

$$
\begin{array}{ccccccc}
 & O & & O & & O & \\
 & | & & | & & | & \\
- O - & Si & - O - & Si & - O - & Si & - O - \\
 & | & & | & & | & \\
 & O & & CH_3 & & O & \\
 & | & & | & & | & \\
- O - & Si & - O - & Si & - O - & Si & - O - \\
 & | & & | & & | & \\
 & O & & O & & O & \\
\end{array}
$$

The more methyl groups one adds to the lattice, the more bridges are broken and the weaker the lattice grows. Eventually, the lattice falls apart into separate fragments, which may be straight chains of silicon-oxygen combinations, branched chains, or rings.

A typical "methyl silicone" would be the following:

$$
\begin{array}{ccccccccc}
CH_3 & & CH_3 & & CH_3 & & CH_3 & & CH_3 \\
| & & | & & | & & | & & | \\
CH_3 - Si & - O - & Si & - O - & Si & - O - & Si & - O - & Si - CH_3 \\
| & & | & & | & & | & & | \\
CH_3 & & CH_3 & & CH_3 & & CH_3 & & CH_3 \\
\end{array}
$$

Such methyl silicones are oily, very much as true oils would be (such true oils being composed of chains of carbon atoms, for the most part). The silicon-oxygen chain is more stable than a carbon chain, however. It is more resistant to change through rising temperature or chemical interaction.

Oiliness arises out of a tendency for long-chain molecules to slide past each other only sluggishly. The more sluggishly they do so, the more viscous the oil; the more slowly it pours, for instance.

A liquid that is at once oily and viscous is useful as a lubricant. The oiliness allows two metal surfaces that are moving with respect to each other to do so over a film of molecules moving past each other, rather than in direct contact. This means the motion is relatively silent and frictionless and inflicts no damage to the metal surfaces. If the liquid is also sufficiently viscous, it does not flow out from between the metal surfaces, but remains there, continuing its good work in preventing damage.

Without lubrication, it would be useless to expect any machines that involve moving parts to work for long.

Ordinary lubricating oils tend to grow less viscous with heat. Rising temperatures accelerate the motion of molecules, encourage the long carbon chains to slide past each other more easily, and increase the danger of the oil leaking away from between the metal surfaces.

In addition, rising temperatures accelerate the speed with which ordinary lubricating oils combine with the oxygen of the air (or with other vapors that may be present in the atmosphere). Such chemical combinations may produce corrosive compounds that will rust the metals, or sludges that reduce the oiliness of the compound, or breaks in the carbon chain that reduce the viscosity. In no case does such chemical combination improve the properties of the lubricating oil.

The sluggish motion of the silicone chains past each other is, on the other hand, scarcely affected by temperature. This means that the viscosity of silicones is relatively constant, declining only slightly as temperature rises.

What is more, silicones are far less likely to combine with various chemicals than ordinary lubricating oils are, and are therefore far less likely to develop any of the undesirable attributes that ordinary lubricating oils would at comparable temperatures.

Silicone lubricating oils maintain their useful properties without trouble at temperatures as high as 150° C, and if oxygen is excluded, at temperatures well over 200° C.

There are also times when it is necessary to lubricate surfaces that are moving past each other at very low temperatures. An ordinary lubricating oil that is at proper viscosity at ordinary temperatures rapidly increases its viscosity as the temperature drops, sets hard, and becomes useless. A silicone lubricating oil does not.

To put figures to it, an ordinary lubricating oil may be 1,800 times as viscous at −35° C as at 40° C (a spread of 75 degrees) while a silicone lubricating oil will be only seven times as viscous at −35° C as at 40° C.

During the crisis days of World War II, the usefulness of silicones in the highly necessary field of lubrication came to be realized and that led directly to the zooming importance of the compounds.

Silicone viscosity tends to increase in a predictable way as the length of the silicon-oxygen chain increases, so that silicone lubricating oils can be easily prepared to suit any viscosity quality needed in a particular job.

If the chain is made long enough, the viscosity becomes high enough to produce solid substances with a rubbery quality. This is particularly so if the chains are connected to each other by a few bridges.

If more connecting bridges are built into the silicon, the result is a resinous substance.

All silicones are electrically nonconducting, so that the silicone rubbers and silicone resins can be used as electrical insulators. They are better than ordinary rubbers and resins in that they are more resistant to heat, and are less likely to become brittle, crack, or otherwise develop flaws in their insulating capacity.

If a silicone is produced with just the proper viscosity, it can even be used as a sort of toy. A silicone may be sufficiently viscous to flow very, very slowly, and resist being hurried. The long molecules will continue to slide past each other with dignity, so to speak, and pressure will be of no use.

A ball of such a substance, thrown against a wall, for instance, will deform under the pressure of contact, but will spring back, as though in hot resentment at having been forced to move at more than its stately will. In other words, it will bounce efficiently.

Put it down on a tabletop, however, and given enough time, it will flatten out, accommodating itself to any unevenness of the surface. Knead it between your fingers and it will be as moldable as wax. It was placed on the market as Silly Putty and I well remember being affected by the craze for it several decades ago.

Silicones, like ordinary carbon-chain molecules, do not dissolve in water. Nor do they mix with water in any way and become waterlogged.

This comes in handy when silicones are added, as a film, to the surface of textiles or other materials.

Thus, you can begin with methylchlorsilane (made up of molecules that consist of a silicon atom attached to three methyl groups and a chlorine atom). This will combine with cellulose, which makes up the bulk of any textile. The cellulose contains oxygen-hydrogen groupings, and the hydrogen atom of such a grouping combines with the chlorine atoms of the methylchlorsilane. This means that a silicon atom, with three methyl groups attached, is hooked on to an oxygen atom of the cellulose and stays there reasonably permanently.

The entire textile surface is thus coated with a layer of silicone that is one molecule thick. The layer can't be seen or felt, but the textile, so layered, will repel water.

Nor is it the methyl group only that can be attached to silicon-oxygen chains. Other carbon-containing groups can be attached to the silicon atoms of such chains. The "ethyl group," for instance, which is made up of two carbon atoms and five hydrogen atoms, can be so used.

We might imagine all kinds of carbon-containing groups attached to the silicon-oxygen chain, a whole series of different kinds, all complicated and each attached to the chain in a different place.

Such very complicated silicone molecules (which we can easily erect in imagination) would be the equivalent of proteins and nucleic acids, though not necessarily mimicking them in structure. We don't have such complicated silicones and, as far as I know, no one is trying to manufacture them, but it seems entirely fair to agree that they could exist in theory. And if so, they may be the basis of a kind of silicon life, or, more properly, silicone life.

But if silicone life were possible, why did it not develop on Earth along with carbon life? Even if carbon life were more efficient in the long run and could win out in a competition, might we not expect silicone life to persist in small amounts, or in out-of-the-way environments where, for some reason, it might prove better fitted than carbon life is?

As far as we know, however, no trace of silicone life exists anywhere on Earth. Nor has it ever existed.

This may be because Earth is too cool for silicone life. If we compare silicone molecules with similar carbon-chain molecules, the one outstanding difference is the greater stability of the silicones, their greater resistance to heat and to chemical change.

That is great if that is what we want, but if we are dealing with life, that is *not* what we want. For life, we don't want stability, but ease of chemical change, the constant flow and counterflow of electrons and atoms.

If life means change, then, at every level, silicone molecules would seem to be less alive, potentially, than carbon-chain molecules are.

It may also be that Earth is too watery for silicone life. All the changes that characterize life take place quickly, and with delicate precision, because the molecules of life are immersed in a watery medium with which they readily interact, and in which some of them dissolve.

Carbon life is impossible, as far as we know, except in the presence of water (or of some liquid with waterlike properties, which includes the possession of polar molecules; that is, molecules in which at least some bits of positive and negative electrical charge are separated asymmetrically).

Silicones, however, have a tendency not to interact with water or with any polar liquid, but to interact with nonpolar liquids instead. Even if carbon-containing atom groups of types that interact with water are added to the silicone chain, the result will be a molecule that is less interactive with water than a corresponding carbon-chain molecule would be.

If life implies interaction with water, or a polar liquid in general, then, at every level, silicone molecules would seem to be less alive, potentially, than carbon-chain molecules are.

But what if we imagine a world that is not cool and watery, as Earth is? Suppose we imagine a world with a temperature well above the boiling point of water. In that case, the temperature may be high enough to make silicone molecules sufficiently active to serve as a foundation for life, and there would be no water to support a competing carbon life (in which the heat would destroy the too-active carbon compounds in any case).

To be sure, there would have to be *some* liquid in which the silicone molecules could dissolve or with which they could readily interact, and it may be that the silicones would supply that as well.

Relatively simple silicones might be nonpolar liquids at elevated temperatures of, say, 350° C, and in them might be dissolved, or otherwise dispersed, the complicated molecules that would be the silicone equivalents of proteins and nucleic acids. We might then have silicone life.

We might further imagine the complex silicones being built up at the expense of solar energy out of silica plus simple carbon compounds, this being the province of the equivalent of silicone plants. Once formed, the complex silicones might, after being eaten by the equivalent of silicone animals, have their carbon-containing portions oxidized away to yield chemical energy, leaving over solid silica as a waste product. (Stanley G. Weinbaum pictured a situation something like this in his short story, "A Martian Odyssey" back in 1934.)

We can raise an objection, however. We picture the complicated silicones as possessing very complex carbon-containing side chains. The more complex these are, the more sensitive they will be to high temperatures, surely. The silicon-oxygen chain is a stabilizing influence, but even it must have its limits.

Eventually, the carbon-containing side chains would be unable to survive the high temperatures, and this, I suspect, will be at some point well short of where we can expect life to exist.

But consider this. The intricate carbon chains and carbon rings of living tissue have most of their bonds taken up by hydrogen atoms so that the compounds of life are, in a way, modified "hydrocarbons." This is possible because hydrogen atoms are extremely small and they can take up most or all of the carbon bonds without getting in each other's way. Only one other atom is small enough to do this and that is the fluorine atom. What if we imagine complicated carbon compounds that are modified "fluorocarbons"?

As it happens, the carbon-fluorine bond is stronger than the carbon-hydrogen bond. Fluorocarbons are therefore stabler and more inert than the corresponding hydrocarbons are, and are more capable of withstanding high temperatures.

We might imagine "fluorosilicones," which are more stable and inert than ordinary silicones, and which could survive the heat required to make them undergo the changes we associate with life.

(This is not a brand-new idea with me. I mentioned it in passing in "Not as We Know It," in *View from a Height,* Doubleday, 1963.)

Next comes another objection. Silicones and fluorosilicones are built of silicon atoms, oxygen atoms, carbon atoms, hydrogen atoms, and fluorine atoms. On the hot planets on which silicones or fluorosilicones might form the bases of life, carbon atoms, hydrogen atoms, and fluorine atoms are quite likely to be very rare, if not virtually nonexistent. These molecules tend to exist as compounds that are easily melted and vaporized, and hot worlds, if rather small (as they would be expected to be), would not be able to hold them with their feeble gravitational fields.

The Moon, which reaches fairly high temperatures during its two-week day, is very poor in these volatile atoms, and we can be certain this is true of the still hotter Mercury as well.

Venus is massive enough to retain a thick atmosphere that contains ample carbon atoms in its carbon dioxide molecules, and a lesser supply of hydrogen atoms in its water molecules. It is conceivable that fluorine atoms might be present, too.

And yet are conditions on Venus such that silicones can be naturally produced, even given the existence of raw materials? It seems to me almost certainly not. I suspect that it would be extremely difficult to design a planet with the kind of chemistry that would allow silicones or fluorosilicones to be formed spontaneously and to have a chance of evolving into the complexity necessary for life.

Yet even though I have taken up three chapters to demonstrate that silicon atoms do not serve as a basis for life, as carbon does—the fact is that a form of silicon life is actually in the process of development here on Earth.

Such life, however, is like nothing I have discussed so far, so in the next chapter we will approach the whole matter from a completely new point of view.

VIII

Silicon Life After All

Every occupation has its hazards; and my own particular niche in the literary world includes the risk of developing a reputation for omniscience. I am forever finding myself on the edge of being expected to know everything.

I deny the impeachment with a bashful fervor every chance I get. In fact, I have a settled routine for the end of every speech, when the time arrives for questions from the audience. I say, "You may ask me anything at all, for I can answer all questions, if 'I don't know' is counted as one possible answer."

Does it help? —No, it doesn't.

In the May 24, 1982, issue of *New York* magazine answers were presented for their "Competition 44," in which the readers had been asked for quotations that were considered humorously inappropriate for the "famous person" to which they were ascribed. Among the honorable mentions was:

" 'I don't know.' —Isaac Asimov."

I'm sure that my science essays are a major factor contributing to this misconception, but I can't help that. I have no intention of ever stopping these essays for any reasons other than mortality.

Let's begin with the notion that an electric current travels easily through some substances but not others. A substance that easily carries a current is an "electrical conductor" or, simply, a "conductor." A substance that does not easily carry a current is, almost inevitably, a "nonconductor."

Not all conductors transmit an electric current with equal ease. Any particular substance offers a certain amount of resistance to the passage of current; and the greater the resistance, the poorer the conductor.

Even if we are dealing with only a single substance, fashioned into a wire, we may expect to have different resistances under different circumstances. The longer the wire, the greater the resistance; the smaller the cross-sectional area of the wire, the greater the resistance. (This would also be true of the more familiar situation of water passing through a hollow pipe, so it shouldn't surprise us.)

Suppose, though, that we compare the resistances of different substances, each of which is made into a wire of the same standard length and cross section and is kept at 0° C. Any difference in resistance would then be entirely due to the intrinsic properties of the substance. It would be the "resistivity" of the substance; and the lower the resistivity, the better the conductor.

Resistivity is measured as so many "ohmmeters," the exact meaning of which is irrelevant right now, and which I won't keep repeating. I shall just give the figures.

Silver is the best conductor known, and has the lowest resistivity— 0.0000000152, or 1.52×10^{-8}. Copper is next with a resistivity of 1.54×10^{-8}. Copper has a resistivity only a little over 1 percent higher than that of silver, and copper is considerably cheaper, so if you care to strip the insulation off a wire used in an electrical appliance, you will find it is copper, and not silver, that forms the wire.

In third place is gold, which is 2.27×10^{-8} (its expense precludes its use), and in fourth place is aluminum with 2.63×10^{-8}.

Aluminum has a resistivity about 70 percent higher than that of copper, but it is so cheap that it is the metal of choice for long-distance transmission of electricity. By making the aluminum wires thicker, their resistance will drop to below that of the usually thinner copper wires; and yet aluminum is sufficiently less dense than copper so that the thick aluminum wires will be less massive than the thin copper wires. Mass for mass, in fact, aluminum is the best conductor.

Most metals are fairly good conductors. Even Nichrome, an alloy of nickel, iron, and chromium, which has an unusually high resistivity for a metal, has one of merely 1×10^{-6}. This is 65 times as high as that of copper, and makes Nichrome a suitable wire for use in toasters, and in heating elements generally. The electric current, forcing its way through the Nichrome, heats it much more than it would heat a copper wire of equivalent size, for the heating effect goes up with resistance, as you might expect.

The reason why metals can conduct electricity comparatively well is that in each metal atom there are usually one or two electrons that are located far out on the atomic outskirts and are therefore loosely held. These electrons can easily drift from atom to atom and it is that which facilitates easy passage of the electric current.

(The movement of electrons is not quite the same as the flow of electricity. The electrons move rather slowly, but the electrical impulse their motion makes possible travels along the wire at the speed of light.)

In substances in which all the electrons are firmly held in place, so that there is little or no drift from one atom to another, electric current flows very slightly. The substance is a nonconductor and the resistivity is high.

Maple wood has a resistivity of 3×10^8; glass one of about 1×10^{12}; sulfur one of about 1×10^{15}, and quartz something like 5×10^{17}. These are outstanding nonconductors.

Quartz has 33 trillion trillion times the resistivity of silver, so that if a quartz filament and a silver wire, of equal length and cross section, were connected to the same electric source, 33 trillion trillion times as much current would pass through the silver in a given unit of time as would pass through the quartz.

Naturally, there are substances that are intermediate in ability to conduct an electric current. The element, germanium, has a resistivity of two, and silicon has one of 30,000.

Silicon has a resistivity that is two trillion times as great as that of silver. On the other hand, quartz has a resistivity that is sixteen trillion times as great as that of silicon.

Silicon (which was the subject of my last three chapters) has, therefore, a resistivity that is about midway between the extremes of conductors and nonconductors. It is an example of a "semiconductor."

In a previous chapter, I explained that of the silicon atom's fourteen electrons, four were on the outskirts and were less tightly held than the rest. In a silicon crystal, however, each of the four outer electrons of a particular silicon atom is paired with one of the outer four of a neighboring atom, and the pair is more tightly held between the two neighbors than a single electron would be. That is why silicon is, at best, only a semiconductor.

The semiconducting property is at a minimum if all the silicon atoms are lined up perfectly in three-dimensional rank and file so that the electrons are held most tightly. In the real Universe, however, crystals are very likely to have imperfections in them, so that somewhere a silicon atom does not have a neighboring atom appropriately placed, and one of its electrons dangles. The occasional dangling electron increases the conductive power of silicon and contributes disproportionately to its semiconducting properties.

If you should desire to have an electric current pass through silicon with reasonable ease, you could help by throwing in a few extra electrons. An easy way of doing this is by deliberately adding an appropriate impurity to the silicon—arsenic, for instance.

Each atom of arsenic has 33 electrons, which are divided into four shells. The innermost shell contains two electrons, the next eight, the next eighteen, and the outermost five. It is these outermost five electrons that are most loosely held.

When the arsenic is added to the silicon, the arsenic atoms tend to take their place in the lattice, each one lining up in some random location where, if the silicon were pure, a silicon atom would have been. Four of the outermost electrons of the arsenic atom pair up with those of neighboring atoms, but the fifth cannot, of course. It remains loosely held, and it drifts.

It may manage to find a place here or there, but only at the cost of displacing another electron, which must then proceed to drift. If one end of such a crystal is attached to the negative pole of a battery and the other to a positive pole, the drifting electrons (each of which is negatively charged) will tend to drift away from the negative pole and toward the positive pole. Such an impure silicon crystal is an "n-type semiconductor," the "n" standing for "negative," which is the charge of the drifting electrons.

Suppose, though, that it is a small impurity of boron that is added to the silicon. Each boron atom has five electrons, two in an inner shell and three in an outer one.

The boron atoms line up with the silicon atoms, and each of the three outer electrons pair up with electrons of the silicon neighbors. There is no fourth electron and in its place there is a "hole."

If you attach such a crystal to the negative and positive poles of a battery, the electrons tend to move, when possible, away from the negative and toward the positive pole. This tendency does little good because ordinarily there is nowhere for the electrons to go, but if an electron has a hole between itself and the positive pole, it moves forward to fill it, and, of course, leaves a hole in the place where it was. Another electron fills that hole, which appears in still a new place, and so on.

As the electrons fill the hole in turn, each moving toward the positive pole, the hole moves steadily in the other direction toward the negative pole. In this way, the hole acts as though it were a positively charged particle so that this type of crystal is termed a "p-type semiconductor," the "p" standing for "positive."

If an n-type semiconductor is attached to a source of alternating current, the excess electrons move in one direction, then the other, then the first, and so on, as the current continually changes direction. The same is true, with the holes moving back and forth, if it is a p-type semiconductor that is in question.

Suppose, though, we have a silicon crystal which has arsenic impurity at one end and boron impurity at the other end. One half of it is n-type and the other half is p-type.

Next imagine that the n-type half is attached to the negative pole of a direct-current battery, while the p-type half is attached to the positive pole. The excess electrons in the n-type half move away from the negative pole to which it is attached and toward the center of the crystal. The holes of the p-type half move away from the positive pole to which it is attached and toward the center of the crystal.

At the center of the crystal, the excess electrons fill the holes and the two imperfections cancel—but new electrons are being added to the crystal at the n-type end, and new holes are being formed at the p-type end as electrons are drawn away. The current continues to pass through indefinitely.

But imagine that the n-type end of the semiconductor is attached to the positive pole of a direct-current battery, and the p-type end is attached to the negative pole. The electrons of the n-type end are attracted to the positive pole to which the end is attached and move to the edge of the crystal away from the center. The holes at the p-type end are attracted to the negative pole and also away from the center. All the electrons and holes move to opposite ends, leaving the main body of the semiconductor free of either, so that an electric current cannot pass through.

An electric current, therefore, can pass through a semiconductor in either direction if the semiconductor is entirely n-type or entirely p-type. If the semiconductor is n-type at one end and p-type at the other, however, an electric current can pass through in one direction, *but not in the other*. Such a semiconductor will allow only half of an alternating current to pass through. A current may enter such a semiconductor alternating, but it emerges direct. A semiconductor that is n-type at one end and p-type at the other is a "rectifier."

Let us next imagine a semiconductor that has three regions: a left end that is n-type, a central region that is p-type, and a right end that is n-type again.

Suppose that the negative pole of a battery is attached to one n-type end, and the positive pole is attached to the other n-type end. The p-type center is attached to a second battery in such a way that it is kept full of holes.

The negative pole pushes the excess electrons of the n-type end to which it is attached away from itself and toward the p-type center. The p-type center attracts these electrons and enhances the flow.

At the other end, the positive pole pulls toward itself the electron excess in the n-type end to which it is attached. The p-type center also pulls at these electrons, however, and inhibits the flow in this half of the crystal.

The p-type center, then, accelerates the flow of electrons on one side of itself, but inhibits it on the other. The overall rate of flow of current can be sharply modified if the extent of positive charge on the central section is shifted.

A small alteration in the charge of the p-type center will result in a large alteration in the overall flow across the semiconductor, and if the charge on the center is made to fluctuate, a similar fluctuation, but a much larger one, is imposed on the semiconductor as a whole. Such a semiconductor is an "amplifier."

Such a three-part semiconductor was first worked out in 1948 and, since it transferred a current across a material that was a resistor (that is, that ordinarily had a high resistance), the new device was called a "transistor." The name was first given it by John R. Pierce (1910–   ), better known to science fiction audiences for the s.f. stories he has written under the name of J. J. Coupling.

Rectifiers and amplifiers are no strangers to the electronics industry. In fact, radios, record players, television sets, computers, and other such devices all depend upon them intensely.

From 1920 to 1950, rectifiers and amplifiers involved the manipulation of streams of electrons forced across a vacuum.

In 1883, the American inventor Thomas Alva Edison (1847–1931) was studying ways to make his filaments last longer in the light bulbs he had invented. He tried including a cold metal filament next to the incandescent one in his evacuated bulb. He noted that an electric current flowed from the hot filament to the cold one.

In 1900, a British physicist, Owen W. Richardson (1879–1959), showed that when a metal wire was heated, electrons tended to boil out of it in a kind of subatomic evaporation, and that this explained the "Edison effect." (Electrons had not yet been discovered at the time of Edison's observation.)

In 1904, the English electrical engineer John A. Fleming (1849–1945) worked with a filament surrounded by a cylindrical piece of metal called a "plate" and placed the whole inside an evacuated container. When the filament was connected to the negative pole of a battery, electrons plunged through it, then out across the vacuum into the plate, so that an electric current passed through the system. Of course, the filament gave off electrons more easily as it grew hotter, so Fleming had to wait some time for the filament to grow hot under the push of electrons before it sprayed them out in sufficiently large quantities to produce a sizable current.

If the filament was connected to the positive pole of a battery, however, electrons were drawn out of the filament and there was nowhere from which replacements could be obtained. They could not be sucked across the battery from a plate that was too cold to yield them. In other words, current could pass in only one direction through the system, which was, therefore, a rectifier.

Fleming called this device a "valve" since it could, in a sense, open or close, permitting or shutting off the electron flow. In the United States, however, all such devices came to be called "tubes," because they were hollow cylinders, and since they came to be best known for their use in radios, they were called "radio tubes."

In 1907, the American inventor Lee de Forest (1873–1961) included a third metal element (the "grid") between the filament and the plate. If a positive charge was placed on that grid, the size of the charge proved to have a disproportionate effect on the flow of electrons between filament and plate, and the device became an amplifier.

Radio tubes worked marvelously well in the control of electron flow, but they did have their little weaknesses.

For instance, each radio tube had to be fairly large, since enough vacuum had to be enclosed for filament, grid, and plate to be far enough apart so that electrons wouldn't jump the gap until encouraged to do so. This meant that radio tubes were relatively expensive, since they had to be manufactured out of considerable material, and had to be evacuated, too.

Since radio tubes were large, any device using them had to be bulky, too, and could not be made smaller than the tubes they contained. As devices grew more and more sophisticated, more and more tubes (each designed to fulfill a special purpose) were required, and bulkiness became more pronounced.

The first electronic computers had to make use of thousands of radio tubes and they were, therefore, perfectly enormous.

Then, too, radio tubes were fragile, since glass is brittle. They were also short-lived, since even the tiniest leak would eventually ruin the vacuum; and if not even the tiniest leak existed to begin with, one would surely develop with time. What's more, since the filaments must be kept at high temperatures all the time the tubes were in action, those filaments would eventually break.

(I remember the time in the early 1950s when I first owned a television set, and had to have what amounted to a "live-in" repairman.

I dread to think what a small proportion of time a computer would be in true working order, when there would never be a time during which some of its tubes would not be going or gone.)

And that's not all. Since the radio-tube filament had to be maintained at a high temperature whenever working, they were energy-consuming. Moreover, since the device did not work until the filament had attained the necessary high temperature, there was always an irritating "warm-up" period. (Those of us past our first youth well remember that.)

The transistor and its allied devices changed all that, correcting every single one of the deficiencies of radio tubes without introducing any new ones. (Of course, we had to wait for some years after the invention of the transistor in 1948, till techniques were developed that would produce materials of the required purity, and that would deliver the required delicacy of "doping" with added impurities, and yet do it all with sufficient efficiency and reliability to keep the prices low.)

Once the necessary techniques were developed, transistors could replace tubes and, to begin with, the vacuum disappeared. Transistors were solid throughout, so that they were called, together with a whole family of similar items, "solid-state devices."

Away went fragility and the possibility of leaks. Transistors were much more rugged than vacuum tubes could possibly be, and much less likely to fail.

What's more, transistors would work at room temperature, so they consumed much less energy, and required no warm-up period.

Most important of all, since no vacuum was required, no bulk was imposed. Small transistors did their work perfectly well, even if there was merely a tiny fraction of a centimeter distance between n-type regions and p-type regions, since the bulk of the material was a far more efficient nonconductor than vacuum was.

This meant that each vacuum tube could be replaced by a far smaller solid-state device. This first entered the general consciousness when computers came to be "transistorized," a term quickly replaced by the far more dramatic "miniaturized."

Computers shrank in size and so did radios. We can slip radios and computers into our pockets now.

Television sets would be miniaturized, too, but we don't want to shrink the picture tubes. That same desire to keep a sizable picture tube limits the shrinkage possibilities of word processors and other forms of computerized television screens.

Over the past quarter century, indeed, the main thrust of computer development has been in the direction of making solid-state devices smaller and smaller, using ever more delicate junctions, and setting up individual transistors that are quite literally microscopic in size.

In the 1970s, the "microchip" came into use, a tiny square of silicon, a couple of millimeters long on each side, upon which thousands of solid-state-controlled electrical circuits could be etched by electron beams.

It is the microchip that has made it possible to squeeze enormously versatile capabilities into a little box. It is the microchip that has made pocket computers not only so small in size, but able to do much more than the giant computers of a generation ago—while costing next to nothing, too, and virtually never requiring repairs.

The microchip has also made the industrial robot possible.

Even the simplest human action, requiring judgment, is so complex that it would be impossible to have a machine do it without including some sort of substitute for that judgment.

Suppose, for instance, you were trying to get a machine to perform the task of tightening nuts (which drove Charlie Chaplin insane in the movie *Modern Times* simply because the task was too simple and repetitive for a human brain to manage it for long).

The task seems so simple that even a human brain of less-than-average capacity can do it without thinking, but consider—

You must see where the nut is; reach it quickly; place a wrench upon it in the proper orientation; turn it quickly to the proper tightness; notice, meanwhile, whether the nut is seated properly on the bolt and correct it if it is not; tell whether it is a defective nut or not, discarding and replacing it if it is, and so on.

By the time you try to build the necessary capacities into an artificial arm in order to get it to duplicate all the things a human being does without any realization of how difficult a task he is performing, you would end up (prior to 1970) with a device that would be totally impractical, and incredibly bulky and expensive—if it could be done at all.

With the coming of the microchip, however, all the necessary details of judgment could be made compact enough and cheap enough to produce useful industrial robots.

Undoubtedly, we can expect this trend to continue. People, who are working on robots these days, are concentrating chiefly in two directions: on supplying them with the equivalent of sight, and on making it possible for them to respond to human speech and to speak in return.

A robot that can see, hear, and speak will certainly move a giant step closer to seeming "alive" and "intelligent."

It is clear, moreover, that what will make a robot seem alive and intelligent will be one thing, and one thing only—the microchip. Without the solid-state devices that lend it its abilities and sense of judgment, a robot would be merely a rather intricate lump of metal, wires, insulation, and so on.

And what is the microchip, stripped to its essentials? —Slightly impure silicon, just as the human brain is essentially slightly impure carbon.

We are now heading, I believe, toward a society composed of two broad types of intelligence, so different in quality as to be noncompeting in any direct sense, each merely supplementing the other. We will have human beings with carbon-based brains, and robots with silicon-based brains. More generally, we will have carbon life and silicon life.

To be sure, the silicon life will be human-made and will be what we call "artificial intelligence," but what difference does that make?

Even if there is no possibility that what we think of as natural silicon life can evolve anywhere in the Universe, there will still be silicon life after all.

And if you stop to think of it, silicon life will be as natural as carbon life is, even if silicon life is "manufactured." After all, there is more than one way to "evolve."

It might well seem to us that the whole function of the Universe was to evolve carbon life; and to a robot, it might well seem that the whole function of carbon life, in turn, is to develop a species capable of devising silicon life. Just as we consider carbon life infinitely superior to the inanimate Universe out of which it arose, a robot might argue that—

But never mind that; I dealt with that point in my story "Reason," which I wrote nearly half a century ago.

ASTRONOMY

IX

The Long Ellipse

I married Janet on November 30, 1973, and a couple of weeks later we had our closest approach to a formal honeymoon. We went off on a three-day cruise on the *Queen Elizabeth 2* in order to see Comet Kohoutek.

As it happened, the sky was uniformly cloudy and it dripped rain constantly so we saw nothing. Even if the sky had been clear we would have seen nothing, for the comet belied its early promise and never grew brighter than a minimal naked-eye visibility. It didn't matter. Under the circumstances, we enjoyed ourselves anyway.

Kohoutek himself was on board, and he was scheduled to give a speech. Janet and I crowded into the theater with the rest.

Janet said, "It's *so* nice to be on a trip where *you* don't have to work and make speeches and we can just sit back and listen."

Just as she said that, the master of ceremonies announced the depressing news that we would not hear Kohoutek after all because he was confined to his cabin with an indisposition.

A soft sigh of disappointment arose from the audience and Janet's heart (which is as soft as warm butter) ached for everyone there. She jumped to her feet and called out, "My husband, Isaac Asimov, will be glad to give a talk on comets, if that's all right with you."

I was horrified, but the audience seemed willing to accept something rather than nothing and in no time at all I found myself on the stage with welcoming applause in my ears. I hastily improvised a talk on comets and afterward said to Janet, "But I thought you had just said it was so nice to be on a trip where I *didn't* have to talk."

"It's different when you volunteer," she explained.

Now we are approaching the time when Halley's Comet,* or, in line with recent habit, Comet Halley, will be appearing in the sky again. Because of the relative position of the comet and Earth when it flashes by, it won't be a very spectacular appearance, but it still deserves an essay, I think.

Comet Halley is, by all odds, the most famous of them all.

It has been appearing in Earth's sky every seventy-five or seventy-six years for an indefinite period of time, certainly since 467 B.C., when there was the first recorded, and surviving, description of it. Suppose we number that appearance as #1.

We don't have records of all the later appearances. For instance, #2 (391 B.C.) and #3 (315 B.C.) are blanks.

The first notable appearance was #7 (11 B.C.), since it is possible that Jesus of Nazareth was born at that time or not long after. There have been suggestions, therefore, that it was Comet Halley that gave rise to the tradition of the "Star of Bethlehem."

Comets, generally, were viewed as portents of disaster, and whenever one appeared in the sky, everyone was sure that something terrible would happen. Nor were they ever disappointed, for something terrible always did happen. Of course, something terrible always happened when a comet wasn't in the sky, too, but no one paid attention to that. To have done so would have been rational, and who wants to be that?

The kind of comet-proclaimed disaster most usually expected was the death of a reigning ruler (though, considering the character and ability of most rulers, it remains a mystery why that should be considered such a disaster).

---

*Short "a" please. I hear entirely too many people giving it a long "a" as though it were "Haley"—an insupportable barbarism.

Thus, in Shakespeare's *Julius Caesar*, Calpurnia warns Caesar of ill omens in the heavens, and says:

When beggars die, there are no comets seen;
The heavens themselves blaze forth the death of princes.

In A.D. 837, Ludwig the Pious ruled the Frankish Empire. He was a well-intentioned, but thoroughly incompetent emperor whose reign was a disaster, for all that he was the son of Charlemagne. At that time, he was fifty-eight years old and had been reigning for twenty-three years, and by the standards of those years it should have surprised no one if, in the ordinary course of nature, he had died at that time.

In that year, however, Comet Halley made appearance #18 and that made Ludwig's death seem imminent. Actually, he did not die for another four years, but this was nevertheless taken as absolute confirmation of the comet's predictive ability.

Appearance #21 came in 1066 just as William of Normandy was preparing to invade England and as Harold of Wessex was preparing to ward off that invasion. This was a situation in which the comet couldn't lose. It was going to be a disaster for one side or the other. As it turned out and as we all know, it was a disaster for Harold, who died at the Battle of Hastings. William went on to conquer England and to establish a line of monarchs who have remained on the throne ever since so the comet was certainly no disaster for him or his line.

Appearance #26 came in 1456, when Comet Halley proved its ability to predict in retrospect. The Ottoman Turks had taken Constantinople in 1453, and this might well have been taken as the kind of catastrophe that threatened all of Christendom (although by then, Constantinople was the merest shadow of what it had once been and its loss had only symbolic value).

Nevertheless, the fall of Constantinople didn't seem an *official* disaster till the comet showed up. *Then* there was panic and an incessant tolling of church bells and intoning of prayers.

The next appearance, #27, was in 1532, when, for the first time, something in addition to panicky outcries greeted it. An Italian astronomer, Girolamo Fracastoro (1483–1553), and an Austrian astronomer, Peter Apian (1495–1552), both noticed that the comet's tail pointed away from the Sun. When the comet passed the Sun, moving from one side to the other, the tail changed direction and *still* pointed away from the Sun. This was the first scientific observation on record in connection with comets.

Appearance #29 came in 1682 and it was then observed by a young English astronomer, Edmund Halley (1656–1742). Halley, who was a good friend of Isaac Newton (1642–1727), was busily engaged in persuading Newton to write a book that would systematize his notions. When the Royal Academy proved reluctant to publish the volume— only the greatest scientific book ever written—because it was likely to be controversial, Halley published it in 1687 at his own expense. (As it happened, he had inherited money in 1684, when his father was found murdered.)

Newton's book included, among other things, his Law of Universal Gravitation, which explained the motions of the planets about the Sun and those of the satellites about the planets.

Might it not explain the motions of the comets as well, remove their apparently unpredictable and erratic appearances, and once and for all banish the silly and baseless panics those appearances engendered?

Halley carefully plotted the path across the sky taken by the comet of 1682 and compared it with the paths taken by other comets, according to those reports that survived. By 1705, he had plotted the course of some two dozen comets and was struck by the fact that the comets of 1456, 1532, 1607, and 1682 had all followed about the same path and had appeared at intervals of about seventy-five years.

For the first time it occurred to someone that different comets might actually be different periodic appearances of the same comet. Halley suggested this concerning these comets—that it was one comet following a fixed orbit about the Sun and that it would appear once more in 1758.

Although Halley lived to the great age of eighty-six, he could not live long enough to see whether his prediction would be verified or not, and he had to endure considerable bad jokes at his expense on the part of those who thought the attempt to predict the arrival of comets to be laughable. As an example, the master satirist Jonathan Swift included a few ill-natured jokes in the third part of *Gulliver's Travels* on the subject.

But Halley was right just the same. On Christmas Day of 1758, a comet was seen approaching and, in early 1759, it blazed in Earth's sky. From then on, it was known as Halley's Comet, or Comet Halley, and this was appearance #30.

Appearance #31 came in 1835. That was the year in which Samuel Langhorne Clemens ("Mark Twain") was born. Toward the end of his life, when family disasters had broken him down into depression and bitterness, he said repeatedly that he had come with the comet and would go with it. He was right. It had blazed in the sky when he was born and it blazed again in its appearance #32 in 1910 when he died.

You might think that once the orbit of at least some comets were worked out, and their appearances shown to be in automatic response to the clockwork-like predictions of the law of gravity, that comets generally would be viewed cold-bloodedly; with admiration, of course, but not with fear.

Not so. It turned out that astronomers felt that Comet Halley would come close enough for Earth to pass through its tail, and at once a howl went up from an incredible number of simple souls to the effect that Earth would be destroyed. At the very least, they insisted, the noxious gases in the comet's tail would poison Earth's atmosphere.

There *were* noxious gases in the comet's tail, but the tail was so rarefied that there would be more such gas in the exhaust of a passing automobile than in a million cubic kilometers of comet tail.

There was, however, no use in trying to explain this because it would involve that horrid old property of rationality. Besides, each ill wind blows good to some people. A number of enterprising rascals made a tidy sum by selling "comet pills" to the simple, pills guaranteed to protect them against all ill effects of the comet. In a way, it was perfectly honest, for those who bought the pills suffered no ill effects from the comet. (Neither did those who didn't buy the pills, of course.)

Now appearance #33 is coming up and I am quite confident that, before the comet arrives, there will be the usual predictions that California will fall into the sea so that a number of people will seek high ground. (In the next chapter, I will go through the appearances of Comet Halley more systematically.)

If a comet, such as Halley's, circles the Sun in obedience to the law of gravity, completing one orbital turn every seventy-five or seventy-six years,† why is it visible for only a short period during this time? Planets, in contrast, are visible throughout their orbits.

For one thing, planets travel about the Sun in orbits that are ellipses of low eccentricity and are close to circles. This means that their distance from the Sun (and from the Earth, too) does not vary greatly as they move through their orbits. If they are visible in part of their orbit, they are therefore visible throughout all of it.

A comet such as Halley, however, moves in an ellipse of high eccentricity, one that is cigar-shaped. At one end of its orbit, it is quite close to the Sun (and to Earth), while at the other, it is very distant indeed. Since it is a small body, even an excellent telescope will reveal it only when it is in that part of the orbit where it moves near the Sun ("perihelion"). It is completely lost to sight when outside this region.

What's more, a comet is a small icy body that, as it approaches the Sun, warms up. The surface ice vaporizes, releasing fine dust that is trapped in the ice. The small comet is therefore surrounded by a huge volume of hazy dust gleaming in the Sun, and the solar wind sweeps that dust out into a long tail. It is all this dust that is prominent, rather than the comet itself, and that dust appears only when the comet is near perihelion. As the comet recedes from the Sun, it freezes again. The halo of dust disappears and only a small solid body is left, which is totally invisible. (A comet that has expended all or most of its gases in previous appearances may have only a rocky core left and may be very inconspicuous even at perihelion.)

Finally, any object in orbit moves more quickly the closer it is to the body it circles. For that reason, a comet moves much more quickly when near the Sun and visible than when far and invisible. This means it remains near the Sun (and visible) for only a brief time and away from the Sun (and invisible) for a long time.

For all these reasons, Comet Halley is visible to the naked eye for only a small portion of its seventy-five-year orbit.

Comet Halley, at perihelion, is only 87,700,000 kilometers from the Sun. At that time, it is closer to the Sun than Venus is. At "aph-

†There is some variation in the interval of return because the influence of planetary attractions on passing comets can slow or speed their motions and thus somewhat change their orbits. There are occasions when a close approach of a comet to a planet—particularly Jupiter—can radically change the cometary orbit.

elion,'' when it is farthest from the Sun, it is 5,280,000,000 kilometers from the Sun, farther away than the planet Neptune. Under those conditions, how does one compare the size of a cometary orbit to those of other Sun-circling objects? A simple rehearsal of distances isn't enough, since in the case of comets that varies so greatly.

We might consider the areas enclosed by the orbits. Then we can get a notion of comparative size regardless of eccentricity.

Thus, the area enclosed by the Moon's orbit as it circles the Earth is about 456,000,000,000 square kilometers and, to avoid the zeroes, let's set that equal to ''1 lunar orbital area'' or ''1 LOA.''

We can compare other satellite orbital areas with that. For instance, the satellite that sweeps out the smallest orbital area as it orbits its planet is Phobos, the inner satellite of Mars. Phobos' orbital area is equal to 0.0006 LOA.

The satellite that sweeps out the largest orbital area is J-IX, the outermost satellite of Jupiter. Its orbital area is 59.5 LOA, or just about 99,000 times that of Phobos. That's a spread of five orders of magnitude among satellites.

But what about the planetary orbital areas?

The smallest one known is that of Mercury. Its orbit sweeps out an area of almost exactly 23,000 LOA, which means that the smallest planetary orbital area is 386 times as large as the largest satellite orbital area. Clearly, the LOA unit is not a convenient one for planetary orbital areas.

The Earth sweeps out an orbit that is equal in area to about 70,000,000,000,000,000 square kilometers so that one Earth orbital area (EOA) is equal to a little over 150,000 LOA.

If we use EOAs as a unit, we can work out the orbital areas without much trouble for each of the planets. It would look like this:

Planet	EOA
Mercury	0.15
Venus	0.52
Earth	1.00
Mars	2.32
Jupiter	27.0
Saturn	91.0

Planet	EOA
Uranus	368
Neptune	900
Pluto	1,560

This is pretty straightforward. The orbital areas are, essentially, the squares of the relative distances of the planets from the Sun.

Now, however, we can deal with the comets on the same basis, remembering to take into account the orbital eccentricities, which are too large to ignore in the case of comets. Consider, for instance, Comet Encke which, of all known comets, has the smallest orbit.

At perihelion, Comet Encke is only 50,600,000 kilometers from the Sun and it is then rather closer than the average distance of Mercury from the Sun. At aphelion, Comet Encke is 612,000,000 kilometers from the Sun, nearly as far from the Sun as Jupiter is. If we work out the orbital area of Comet Encke, it comes to 2.61 EOA.

In other words, Comet Encke sweeps out an orbital area only a little larger than that of Mars. Though in distance from the Sun it may move outward nearly as far as Jupiter, its orbit is merely a fat cigar compared to Jupiter's circle, so the orbital area of Comet Encke is only one tenth that of Jupiter.

And what about Comet Halley now, which comes in as close as Venus to the Sun at one end of its orbit and retires farther than Neptune at the other end?

Its orbital area turns out to be 82.2 EOA, nearly that of Saturn.

Suppose we compare ellipses. Every ellipse has a longest diameter, the "major axis," stretching from perihelion to aphelion through the ellipse's center. It also has a shortest diameter, the "minor axis," passing through the center at right angles to the major axis.

The major axis of Comet Halley is 5,367,800,000 kilometers long, which is 8.1 times as long as that of Comet Encke (where it is a mere 662,600,000 kilometers long). The minor axis of Comet Halley is 1,368,800,000 kilometers, which is 3.9 times the length of that of Comet Encke (where it is 352,500,000 kilometers long).

Notice that Comet Halley has a major axis that is 3.92 times as long as its minor axis, whereas Comet Encke has a major axis that is only 1.88 times as long as its minor axis. The proportions of the orbit are that of a slimmer ellipse, a longer and thinner cigar than that of the orbit of Comet Encke. This is just another way of saying that Comet Halley has a larger orbital eccentricity than Comet Encke has. The orbital eccentricity of Comet Encke is 0.847, while that of Comet Halley is 0.967.

Although Comet Halley has an orbit that stretches out beyond Neptune and although it requires seventy-five years to orbit the Sun, Comet Halley is still a "short-term comet." Relatively speaking, it hugs the Sun and circles it quickly.

There are comets that are far more distant from the Sun than Comet Halley is; comets that circle the Sun at distances of a light-year or more and take a million years or more to complete an orbit. We haven't ever seen these far-distant comets, but astronomers are reasonably certain they are there (see "Stepping Stones to the Stars," in *Fact and Fancy*, Doubleday, 1962).

Now, of course, we know of a comet that, while it is not quite one of those far distant ones, certainly has an orbit far greater than that of Comet Halley.

It is none other than Comet Kohoutek. It may have been "the comet that failed" because it never became as bright as astronomers had first hoped, but in a way that was not the fault of the astronomers. Comet Kohoutek had been spied approaching (by Kohoutek, whose place on the platform I had taken on the *QE 2*) while it was still beyond Jupiter and that indicated a large comet. No other comet had ever been first seen at such a distance.

If Comet Kohoutek had been of similar constitution to that of Comet Halley—mostly icy material—it would have formed an enormous haze that would have swept out into a formidable tail and it would have been far brighter than Comet Halley. Unfortunately, Comet Kohoutek must have been a rather rocky one so that, as it approached perihelion, too little ice was present to vaporize and produce much of a haze. For that reason, Comet Kohoutek turned out disappointingly dim for its size.

Nevertheless, it was a remarkable comet for it turned out to have an enormous orbit, the largest orbit of any known and observed object in the Solar System.

At its closest to the Sun, Comet Kohoutek is at a distance from it of only 37,600,000 kilometers, so that it is closer to the Sun than Mercury is. It recedes, however, to a distance of about 1/18 of a light-year at aphelion, 75 times as far as Pluto at its farthest from the Sun.

Even the minor axis is 6,578,000,000 kilometers long, which is a mighty distance. It means that the ellipse marked out by the motion of Comet Kohoutek about the Sun is wider, at its widest, than the full width of the orbit of Uranus.

This long minor axis shrinks, however, in comparison with the still more enormous length of the major axis—which is 538,200,000,000 kilometers.

The major axis of the ellipse that makes up the orbit of Comet Kohoutek is 81.8 times as long as that of Comet Halley, while the minor axis of Comet Kohoutek is only about five times as long as that of Comet Halley. This makes it obvious that the orbital eccentricity of Comet Kohoutek is far greater than that of Comet Halley. The orbital eccentricity of Comet Kohoutek is 0.99993, far greater than that measured for any other body in the Solar System.

The next question is: What is the orbital area of Comet Kohoutek? —The answer is: About 120,000 EOU, or about 77 times the orbital area of Pluto. That is enormous—though even so, it is but a small fraction of the orbital areas of the truly distant comets that circle the Sun at light-year distances.

Comet Kohoutek affects the Sun by swinging in and out to such enormous extremes. If we assume that Comet Kohoutek is a solid body of rock and ice about 10 kilometers across, it would, in that case, have a mass equal to one or two quadrillionths that of the Sun. As Comet Kohoutek swings about in its elliptical orbit about the center of gravity of the Sun / comet system, the center of the Sun must do the same in such a way that the comet and the solar center remain always on opposite sides of the center of gravity. Naturally, the motion of the Sun and the comet must be in inverse proportion to their respective masses, so that if the Sun is a quadrillion or two times as massive as the comet, it moves that many times less in distance.

Even so, as Comet Kohoutek moves out to a distance of 1/18 of a light-year in one direction and then back, the Sun's center moves 10 to 20 centimeters in the other direction and then back. (Naturally, this motion is utterly masked by the much huger motions of the Sun in balancing the much more massive planetary bodies—especially Jupiter—even though these move through smaller distances.)

One more thing: How long does it take Comet Kohoutek to make one swing about its orbit? Using Kepler's Third Law for that, we find that Comet Kohoutek visits the neighborhood of the Sun once every 216,500 years.

This explains why the astronomers were caught by surprise by the dimness of Comet Kohoutek. They couldn't be guided by the similarly disappointing performance on its previous appearance since, when it previously appeared, there were no human beings other than early Neanderthalers to observe it.

And next time it appears, who knows if any human beings will be here to greet it, or whether, if there are, any of the records from 1973 will have survived.

Imagine, though, that there are living things on Comet Kohoutek intelligent enough to be aware of a star in the sky, one that is much brighter than the others, and yet is still only a star.

For many thousands of years it would remain "only a star," not altering perceptibly in brightness. And then would come the time when the cometary astronomers might detect that actually the star seemed to be brightening very slightly—*very* slightly. The brightening would continue. Indeed it would begin to seem to be proceeding at a very slightly accelerating pace and the rate of acceleration would itself be accelerating.

Eventually, the star would come to seem a tiny glowing globe in the sky and would be expanding wildly, then bloating madly into a blaze of impossible heat and light.

If we imagine the living things as surviving, they would see that ball of heat and light reach a maximum, then shrink rapidly, continue to shrink more slowly and still more slowly, fading to a bright star again. The star would slowly dim for a hundred thousand years, then as slowly brighten for a hundred thousand years until, once again, there would come that mad flash of heat and light.

If any of you would like to write a science fiction story set on a planet with an orbit like that—be my guest.

X

Change of Time and State

In our time-bound society, we expect things to happen regularly, and in accordance with the insistence of the calendar and wristwatch.

I belong, for instance, to a group that meets regularly each Tuesday for lunch, and a couple of weeks ago, there were comments made concerning the fact that a particular member had missed a few meetings. The errant member advanced excuses which were dismissed as insufficient in a more or less good-natured way.

At which I saw a chance of advancing both my virtue and my reputation for gallantry by saying, "The only time *I* would miss a meeting would be if the young lady in bed with me simply *refused* to let me leave."

Whereupon one of the gentlemen at the table, a Joe Coggins, promptly said, "Which accounts for Isaac's perfect attendance record," and I was wiped out.

The laughter at my expense was unanimous, for even I had to laugh.

Regularity has always been valued, even before the day of clocks. Something that happened when it was supposed to happen offered no shocks, no possibilities of unpleasant surprises.

The planets, which *seem* to move erratically against the backgrounds of the stars, were carefully studied until those movements were reduced to rule and could be predicted. That was the justification of ancient astronomy, since by knowing how the various planetary positions would relate to each other, astronomers could judge in advance their influence on the Earth, and thus predict events. (We call that sort of thing astrology now, but never mind.)

But then, every once in a while, there came along a comet; appearing from nowhere; vanishing into nowhere. There was no way of predicting its coming and going, and it could only be taken as a warning from above that something unusual was about to happen.

Thus, at the very start of *Henry VI*, Part One, English nobles are standing around the bier of the conquering Henry V, and Shakespeare has the Duke of Bedford intone:

> *Hung be the heavens with black, yield day to night!*
> *Comets, importing change of time and states,*
> *Brandish your crystal tresses in the sky,*
> *And with them scourge the bad revolting stars*
> *That have consented unto Henry's death!*

In other words, the presence of a comet in the sky means that the conditions of life (time) and national and international affairs (states) will change.

In 1705, the English astronomer Edmund Halley (1656–1742) insisted that comets were regular phenomena, circling the Sun as the planets did, but in highly elliptical orbits so that they were seen only near the time of their "perihelia," when they were close to the Sun and the Earth.

The comet whose orbit he calculated, and whose return he predicted, has been known ever since as "Halley's Comet," or, in line with recent habit, "Comet Halley." It *did* return as predicted, and then twice more. Now, in 1986, we expect it still again. (I discussed this comet in the previous chapter, but now let's be more systematic.)

The taming and regularization of comets has not, however, altered the expectations of the unsophisticated. Each time Comet Halley returns (or each time any other comet shows up spectacularly, for that matter) there is fright and panic. After all, just because Comet Halley returns periodically, and just because its return is expected and predicted, doesn't mean that it won't bring on something important and probably uncomfortable. Maybe such events are scheduled, by Providence, in an orderly and periodic manner.

Let's see, then—

Comet Halley completes one revolution about the Sun in seventy-six years or so. The period of revolution is not quite fixed because the comet is subject to the gravitational pull of the planets it passes (particularly to the pull of giant Jupiter). Since at each passage toward the Sun, and then, again, away from it, the planets are at different spots in their orbits, the pattern of gravitational pull is never quite the same, twice running. The period can therefore be as short as seventy-four years, or as long as seventy-nine years.

The earliest report of a comet that seems to have been Comet Halley came in 467 B.C. Counting that appearance, Comet Halley has now been in the sky thirty-two times in the last twenty-four and a half centuries. In 1986, there will be a thirty-third appearance.

We could run through them, and see what "changes of time and state" have taken place with each appearance—if any.

1. *467* B.C. Persians and Greeks have been fighting for a generation and Comet Halley now gleams in the sky to mark the end. In 466, the Athenian Navy defeats the Persians in a great battle off the coast of Asia Minor and the long war is over. Comet Halley also marks a beginning, for in that same year, the democratic party wins control of Athens, and that city begins its Golden Age—perhaps the greatest flowering of genius in one small area over one short period the world has ever seen.

2. *391* B.C. The city of Rome, in central Italy, was very slowly becoming more important. It had been founded in 753 B.C., and had become a republic in 509 B.C. It had been gradually establishing its domination over the neighboring towns in Latium and Etruria. Then came Comet Halley in the sky, and with it the barbarian Gauls from the north. In 390 B.C., the Gauls defeated the Romans north of the city and swept in to occupy Rome itself. The Gauls were finally bought off but the Romans were badly shaken. Apparently, though, it put them into a no-nonsense mood, for after the occupation they moved toward greatness much more rapidly than before.

3. *315* B.C. Between 334 B.C. and his death in 323 B.C., Alexander the Great had swooped like a raging fire over western Asia, conquering the vast Persian Empire in a series of incredible victories. Alexander's empire, however, wasn't permanent, but fell apart immediately after his death as his generals quarreled over the fragments. With Comet Halley shining down, it became clear that there was no chance of the empire being reunited. Antigonus Monophthalmos, who was the one general who was unwilling to settle for less than all, was defeated in 312 B.C., and though he struggled on for another dozen years, it was clear that the empire had been fragmented into the three major Hellenistic kingdoms: Egypt under the Ptolemies; Asia under the Seleucids, and Macedonia under the Antigonids.

4. *240* B.C. The Hellenistic kingdoms fought each other continually, with no clear-cut victory for any one of them, merely a steadily increasing overall exhaustion. By 240 B.C., when Comet Halley was once again in the sky, it was clear that the Hellenistic kingdoms were declining and that other nations were on the rise. About 240 B.C., Arsaces I was establishing his power in Parthia, an eastern province of what had once been the Persian Empire. What's more, in 241 B.C., Rome, which controlled all of Italy, had defeated Carthage (which controlled North Africa) in the First Punic War. Rome was now dominant in the western Mediterranean. Comet Halley thus marked the rise of powers East and West, which, between them, would destroy the Hellenistic kingdoms.

5. *163* B.C. When Comet Halley returned, it was to mark the fact that Rome had defeated Carthage a second time, in 201 B.C., and had gone on to destroy Macedonia and reduce the Seleucid kingdom and the Ptolemies of Egypt to puppets. Rome had just established a clear dominance over the whole Mediterranean by 163 B.C. and was entering its greatest period. Meanwhile, in Judea, a small province of the Seleucid kingdom, the Jews had risen in revolt. A combination of inspired leadership by Judas Maccabeus and internal squabbling among the Seleucid royal family led to Jewish control of Jerusalem in 165 B.C. and to a *de facto* recognition of Jewish independence by the Seleucids in 163 B.C. Comet Halley blazed, then, over a newly Roman Mediterranean and a newly Jewish Judea, and the time would come when they would interact with important consequences.

6. *87* B.C. The Roman governmental system, which had suited a small city fighting for control of a province, was breaking down under the stresses of attempting to govern a large empire of diverse peoples, languages, and customs. Internal infighting between Roman politicians grew steadily more deadly, especially since each side tended to be backed by one general or another, so that political quarrels degenerated into civil war. General Marius favored the democratic side; General Sulla, the aristocratic side. In 87 B.C., Comet Halley returned and illuminated a crucial moment, for in that year Sulla and his army forced their way into the city of Rome and slaughtered some of the more radical politicians. It was not the Gauls who occupied Rome this time, but a Roman general. The portent of Comet Halley was clear. No external enemy might be able to stand up to Rome, but Rome would be torn apart by divisions within.

7. *12* B.C. Comet Halley returned to find that Rome had fought its way through a whole series of civil wars and had survived and had even expanded and grown stronger. It had become the Roman Empire and, under its first emperor, Augustus, it lay in profound peace except for local fighting along its northern borders. Somewhere about this time, Jesus is supposed to have been born in Bethlehem. The exact year of his birth is not known, but some people think it may have been 12 B.C. and maintain that Comet Halley is the "star of Bethlehem." If so, Comet Halley imported a change at this appearance that, to many people, was the central change of history.

8. *66.* The Roman Empire was still largely at peace at Comet Halley's next return, and was ruled by Nero. One area of discontent was Judea, however. It dreamed of a Messiah and wished to emulate the old Maccabean fight and free itself of Rome. In 66, with Comet Halley overhead, Judea broke into revolt. It was defeated in a bloody four-year struggle. Jerusalem was sacked and the Temple destroyed. The fate of a small province didn't seem to matter much, but the new group of Christians had held aloof from the struggle and it lost all standing with Jews in consequence. This meant the Christians were no longer a Jewish sect, but became an independent religion of increasingly Greco-Roman cultural content, and this, in turn, had a profound effect on future history.

9. *141*. The Roman Empire, by the next return of Comet Halley, had lived through a plateau of peace and prosperity climaxed by the almost eventless reign of Antoninus Pius, who became emperor in 138. Comet Halley now shone down on the culmination of Mediterranean history. All the struggles of the Greeks and Romans among themselves and with others had ended with the Mediterranean world united under an enlightened and civilized government. It was something the region had never seen before and would never see again. Comet Halley marked that climax. The upward climb had ceased, the downward slide would begin.

10. *218*. The happy period of the good emperors was long gone by the next return. After some disorders, Septimius Severus placed the empire under strong rule again. In 217, however, his son, Caracalla, was assassinated, and Comet Halley shone down upon the beginning of a long period of anarchy during which the empire nearly went under. Comet Halley had presided over the peak on its previous appearance, and now it marked the beginning of a trough.

11. *295*. The period of anarchy came to an end in 284 with the coming to power of Diocletian, the first strong emperor to have a fairly long and stable reign since Septimius Severus. Diocletian set about reorganizing the Imperial Government, and made it into an oriental monarchy. The vestigial remnants of old Rome disappeared and in 295 Comet Halley was presiding over the coming of a revised empire in which, from now on, the eastern half would be dominant. It was almost like a return to Hellenistic times.

12. *374*. The reforms of Diocletian kept the Roman Empire going, but the relief was only temporary and time was running out when Comet Halley was in the sky again. The Huns were on the march, pouring out of Asia and across the Ukrainian steppes, driving the Goths (a Germanic tribe) before them. In 376, some of the Goths, seeking refuge, crossed the Danube into Roman territory. The Romans mistreated them and, in 378, the Roman legions were defeated and destroyed by the Gothic cavalry at the Battle of Adrianople. A new age had dawned, and cavalry would dominate the battlefield for a thousand years. Comet Halley was presiding over the fall of the old Roman Empire and the rise of the German tribes.

13. *451.* By the time Comet Halley returned again, several of the western provinces of the Roman Empire were under the direct control of German warlords, and the Huns were stronger than ever. Under their ruler, Attila, a Hunnic empire, stretching from the Caspian Sea to the Rhine River, existed. In 451, with Comet Halley in the sky, Attila penetrated into Gaul, the farthest westward any of the central Asian nomads were to penetrate either before or after. At the battle of Chalons, however, combined Roman and German forces fought Attila to a standstill. Two years later, he died and the Hunnic empire fell apart. Comet Halley thus presided over the cresting of the central Asian flood.

14. *530.* By the time Comet Halley returned, the Roman Empire in the West had fallen and the provinces were all German-controlled. The greatest of the new leaders was Theodoric the Ostrogoth, who ruled Italy in enlightened fashion and labored to preserve Roman culture. However, Theodoric died in 526 and the next year Justinian I became the East Roman Emperor. Justinian planned to reconquer the West, and in 533, his general, Belisarius, sailed west to begin a process that devastated Italy, destroyed the Ostrogoths, yet did *not* really restore the empire. The West was left to the untouched Franks, the most barbarous of the Germanic tribes. In this way, Comet Halley shone down on the beginning of the campaigns that established the Dark Ages.

15. *607.* The Roman Empire in the East remained strong and intact, but in 607, as Comet Halley shone in the sky, the Persians, under Chrosroes II, began their last and most successful war against the Romans. At the same time, in Arabia, a young merchant named Muhammad was developing a new religion based on his version of Judaism and Christianity. The Persian-Roman war succeeded in utterly exhausting both combatants, and the new religion would take over all of the Persian Empire and more than half of the Eastern Roman Empire against much-diminished capacity for resistance. Thus, Comet Halley beamed down upon the beginning of Islam and a newly shrunken remnant of the Roman Empire, now called the "Byzantine Empire."

16. *684.* In an amazing sweep of success, the Arabic followers of Islam surged out of Arabia following the death of Muhammad and took over Persia, Babylonia, Syria, Egypt, and North Africa. They

were ready now to take over Constantinople itself, then sweep through Europe and consolidate their hold on the entire western world. They laid siege to Constantinople even as the barbarian Bulgars swept down the Balkans to approach the city from the land side. Constantinople held, however, defeating the Arabs with Greek fire in 677. In 685, after Comet Halley had appeared, Justinian II succeeded to the throne, a cruel but energetic ruler who defeated the Bulgars. Comet Halley presided over the survival of the Byzantine Empire as the shield of Europe against Islam.

17. *760*. Islam continued to expand in lesser ways and took Spain, for instance, in 711. In 750, the Abbassid Caliphate was established, with its capital at Baghdad, and ruled over all of Islam except for Spain and Morocco. By 760, with Comet Halley in the sky, the caliphate was firmly established, and, for a period of time, Islam was at its peak, peaceful, united, and powerful beyond challenge. As Comet Halley had shone down upon the peak of the Roman Empire eight appearances ago, it now shone down upon that of the Islamic Empire.

18. *837*. In the West, the Frankish Empire reached its peak under Charlemagne, who had died in 814. His successor, Louis the Pious, reigned over an intact empire, but he was weak, and had four sons among whom he intended to divide the realm. There were civil wars over the matter, but in 838, the final plan for division was concluded. Comet Halley thus presided over a division that was never to be healed, and the history of Europe forever after was that of a multiplicity of ever-warring nations. What's more, Comet Halley's appearance heralded new invasions from without. The Vikings, from the north, launched their most dangerous raids soon after 837, as did the Magyars from the east, while the Arabs from North Africa were invading Sicily and launching incursions into Italy.

19. *912*. The last major incursion of Viking forces into Frankish territory was that of the Northmen, or "Normans," under Hrolf. In 912, with Comet Halley in the sky, Rollo accepted Christianity and was awarded a section of the Channel coast to rule. The region has been called "Normandy" ever since. Thus, Comet Halley presided over the birth of a new state that was to play an important role indeed in European history.

20. *989*. Comet Halley, on its return, presided over the shaping of modern Europe. The descendants of Charlemagne had come to their final end, and in what is now called France, a new line, in the person of Hugh Capet, came to the throne in 987. His descendants will rule for nine centuries. In 989, Prince Vladimir of Kiev is converted to Christianity and this introduces the appearance of Russia as a European nation. Comet Halley is presiding over the end of the Dark Ages as it had presided over the beginning six appearances ago.

21. *1066*. Normandy, which was formed two appearances ago, had by now become the best-governed and most powerful realm in western Europe under its extremely capable Duke William. Normans had already drifted to the Mediterranean, where they took over Sicily and southern Italy. William, however, planned an invasion of England, just across the Channel. Comet Halley appeared even as the fleet was being prepared, and before the year was over, he had won the key Battle of Hastings and had become William the Conqueror. Comet Halley thus saw the formation of a Norman England, which, in time, was to outdo both Rome and Islam.

22. *1145*. Reviving Europe attempted its first general offensive against the non-European world in 1096, when armies poured eastward in a Crusade to retake Jerusalem. The armies were ill organized, ill equipped, ill led, but they were full of the valor of ignorance and they faced a fragmented enemy. They took Jerusalem in 1099 and established Christian kingdoms in the Holy Land. Slowly, however, Islam rallied against the invader and, in 1144, scored their first major success when they retook Edessa in the northeastern corner of the European conquest. Comet Halley shone down upon calls for a second Crusade, which, however, was to prove a fiasco. The crusading movement continued, but in the long run it failed, and the appearance of Comet Halley marks virtually the exact moment when that failure became evident.

23. *1222*. Europe was not yet ready, by any means, to rule the world. As Comet Halley returned, a new menace from Asia had arisen that, for a while, was greater than any that had preceded it or was to follow it. In 1162, a Mongol named Temujin had been born. By 1206, he ruled the tribes of central Asia under the name of Genghis Khan. He forged them into a fighting army trained in brilliant new tactics

that capitalized on mobility, surprise, and relentless shock. In a dozen years, he took northern China and swept across western Asia. In 1222, with Comet Halley in the sky, a Mongol army made its first appearance in Europe and the next year that army inflicted a resounding defeat on the Russians. The Mongols then left, but they were to return. Comet Halley presided over the beginning of the disaster.

24. *1301*. The Mongols came again and won victory after victory, but retired, undefeated, to elect a new monarch. Russia remained in their grasp and its entire future history was distorted as a result. When Comet Halley returned, that episode was over, and other significant events were taking place. The European knights, who had ruled the battlefield for centuries, rode against the rebelling burghers of Flanders. The knights were filled with contempt for the low-born varlets. The low-born varlets, however, had pikes and chose their ground well. They destroyed the French horsemen at the Battle of Courtrai in 1302. Meanwhile, Pope Boniface VIII capped the increasing power of the papacy by laying claim in 1302 to supreme rule over the kings of Christendom. Philip IV of France thought otherwise and sent agents to manhandle the Pope (who soon died) and then established a papacy that served as a French puppet. Thus, Comet Halley, on this appearance, presided over the beginning of the end of the feudal army, and of the all-powerful papacy as well, and, therefore, over the beginning of the end of the Middle Ages.

25. *1378*. After Boniface VIII, the papacy is established in Avignon, a city in southeastern France. In 1378, with Comet Halley again in the sky, a Pope reestablished himself in Rome. The French cardinals, however, unwilling to leave Avignon, selected a Pope of their own. This started the "Great Schism," which lasted for forty years, and which provided Europe with the spectacle of rival popes anathematizing and excommunicating each other while nations chose sides according to their secular interests. The prestige of the papacy was destroyed and the groundwork was laid for changes that would forever destroy the religious unity of Europe, as the political unity had been destroyed seven appearances ago.

26. *1456*. When Comet Halley reappeared, it was to find that the Ottoman Turks were now the cutting edge of Islam. Since 1300, they had been expanding their power in Asia Minor, and in 1352 they made

their first appearance on the European side of the Hellespont. In 1453, they took Constantinople itself, thus putting an end to the Roman dominions twenty-two centuries after the founding of Rome. In 1456, with Comet Halley in the sky, the Ottoman Turks took Athens and laid siege to Belgrade. Western Europe was well aware of the new threat from Asia that Comet Halley heralded.

27. *1531*. The Ottoman Empire reached its peak under Suleiman the Magnificent, who conquered Hungary and who, in 1529, laid siege to Vienna. Vienna held out, however, and the Ottoman Turks retreated to Budapest. Meanwhile, Columbus has discovered the American continents and, as Comet Halley shone down on the newly liberated Vienna, Spanish conquistadores, having conquered the Aztecs of Mexico, are leaving for Peru, where they will destroy the Inca Empire within two years. Thus, Comet Halley presides over a Europe that has managed to stop the Ottoman advance and, at the same time, to establish itself far across the ocean. Europe is on the threshold of world domination.

28. *1607*. In 1607, at the return of Comet Halley, a group of Englishmen found Jamestown in a region they call Virginia. It is to be the first permanent English colony established on the eastern coast of North America, and it is the start of a series of developments that will end with the establishment of the United States of America, which will, in days to come, dominate Europe for a period of time.

29. *1682*. With Comet Halley again in the sky, Fedor III of Russia died, and was succeeded by his two sons as co-emperors. The younger was Peter I, who, in time to come, would be called "Peter the Great," and would, with titanic energy, drag Russia out of the twilight of its Mongol-dominated past and into the sunlight of western European advance. Russia would remain Western in orientation and, as a result of Peter's labors, would someday dispute the world with the United States.

30. *1759*. Europe dominated the world by the next return of Comet Halley, but which European nation would have the lion's share? Spain and Portugal were first off the mark, but they had decayed. The Netherlands had made a valiant try but it was too small. England (now Great Britain) and France were the remaining candidates and, in 1756, the decisive "Seven Years War" between them started. (Prussia, Aus-

tria, and Russia also participated.) The turning point came in 1759 when, with Comet Halley in the sky, Great Britain took Canada, gained control of India, and proved itself undisputed master of the seas. Comet Halley shone down upon the true foundation of the British Empire, which would dominate the rest of the globe for nearly two centuries.

31. *1835*. Great Britain, world leader, was changing peacefully by the next return of Comet Halley. In 1832, a reform bill was forced through Parliament that rationalized representation in that body, extending the electorate and beginning the process of broadening the franchise to the population generally. Victoria reached the throne in 1837. In the United States, the first whiff of a split between North and South came with the nullification crisis of 1832, which was eventually resolved in favor of the Union. The battle lines were drawn, however, and in the end, the franchise would be extended to the freed slaves. In both nations, movement toward egalitarian doctrine took a firm course forward, with Comet Halley in the sky.

32. *1910*. Edward VII of Great Britain, oldest son of Queen Victoria, died in 1910, and at his funeral there was an outpouring of the crowned heads of Europe for the last time. In 1914, World War I was to begin. It would destroy many of the ancient monarchies, and establish a new and more dangerous world. Once again, Comet Halley was importing the change of time and states.

33. *1986*. ?

Impressive, isn't it? Maybe there's something to astrology after all.

No, there isn't. This is just a tribute (if you'll excuse the immodesty) to my ingenuity. Give me any list of thirty-three dates from 700 B.C. on, spaced regularly or irregularly, and give me a little time to think, and I will undertake to draw up a similar list of crucial events, sounding just as good. Given fifty such lists (especially if we include oriental history, technological advance, cultural events, and so on), it would be easy to set up fifty interpretations and it would be hard to choose a particular one as best.

Human history is sufficiently rich, and the currents sufficiently full of branch points, to make this possible, and that is one of the reasons why my imaginary science of psychohistory is going to be so hard to develop.

XI

Whatzisname's Orbit

I have just returned from the Institute on Man and Science at Rensselaerville, New York, where, for the ninth successive year, I helped conduct a seminar on a science-fictional subject. This time it was on space treaties.

How, for instance, do we regulate the use of the limited space in a geosynchronous orbit, considering that that is where it would make the most sense to place a solar power station?

Speaking was my good friend Mark Chartrand, who is now the head of the National Space Institute. On several occasions, he referred to the geosynchronous orbit as the "Clark orbit."

I was puzzled, and finally I spoke up. "Why the Clark orbit?" I asked. "Who's Clark?"

Chartrand stared at me for a moment, and said, "The reference is to Arthur C. Clarke. Surely you have heard of him, Isaac."

There was loud laughter, naturally, and when it died down, I said, indignantly, "Well, how the heck was I supposed to know you were referring to Arthur? You didn't pronounce the silent 'e' in his name."

Would you believe that no one considered that an adequate excuse?

The point is (and it is something I very well knew) that, way back in 1945, Arthur Clarke had discussed the possibility of placing communications satellites in orbit, and had described the particular usefulness of having them in geosynchronous orbit. This was the first time the point had been made, I believe, and so "Clarke orbit" is a perfectly justified term.

To make up for my failure to recognize Arthur's name when I heard it, let's go into the Clarke orbit in detail.

Suppose we consider various objects revolving about the Earth at various distances from its center. The farther an object is from Earth, the longer the orbit through which it must sweep, and, at the same time, the more slowly it must travel, since the intensity of Earth's gravitational field decreases with distance.

The period of revolution, depending, as it does, upon both the length of the orbit and upon the orbital speed, increases with distance in a way that is slightly complicated.

Thus, imagine a satellite skimming about the Earth no more than 150 kilometers above its surface, or (which is the same thing) about 6,528 kilometers from its center. Its period of revolution is just about 87 minutes.*

The Moon, on the other hand, revolves about the Earth at a mean (that is, average) distance of 384,401 kilometers from Earth (center to center). Its "sidereal" period of revolution (that is, its revolution relative to the stars, which is the closest we can get to the concept of its "real" revolution) is 27.32 days. The Moon is 58.19 times as far from Earth's center as the satellite is, but the Moon's period of revolution is 452 times as long as the satellite's.

The period, it would seem, lengthens more quickly than the distance does, but less quickly than the square of the distance does. We can put this mathematically by calling the ratio of the periods of revolution as P and the ratio of distances as D, and saying that $P > D^1$ and $P < D^2$, where $>$ means "is greater than" and $<$ means "is less than." As it turns out, $P = D^{1.5}$.

An exponent of 1.5 means that to get the period of the farther object you must take the cube of the ratio of the distances and then take the

---

* All moving objects in this essay are assumed to be moving in a prograde direction (west to east), the direction of Earth's rotation.

square root of the result. Thus, the Moon is 58.9 times as far from the Earth as the satellite is. Therefore, let us take the cube of that ratio—58.9 × 58.9 × 58.9 = 204,336—and then take the square root of that, which is 452. That is the ratio of the periods of revolution. If you multiply the satellite's period by 452, you will get the Moon's period. Or you can begin with the sidereal period of the Moon, divide that by 452, and get the period of the asteroid. Or, beginning with the ratio of the periods, you can get the ratio of the distances.

All this is Kepler's Third Law, and now we can forget about the mathematics. I'll do the calculating and you can take my word for it.†

The Earth turns on its axis, relative to the stars (the "sidereal day"), in 23 hours 56 minutes. The Earth's sidereal day is longer than the period of revolution of the satellite skimming its surface, and shorter than the period of revolution of the Moon.

As we imagine a series of objects revolving about the Earth in orbits that are farther and farther away from the planet's center, the period of revolution will grow longer and longer, and at some distance between that of the satellite (where the period is too short) and that of the Moon (where it is too long) there will be a place where a satellite will have a sidereal period of revolution exactly equal to that of Earth's sidereal period of rotation.

Such a satellite is moving in a geosynchronous orbit, "geosynchronous" being from Greek words meaning "moving in time with the Earth."

Using Kepler's Third Law, we can find out exactly where a satellite must be in order to be in a geosynchronous orbit.

It turns out that a satellite, revolving about the Earth at a mean distance of 42,298 kilometers from Earth's center, will revolve in precisely one sidereal day. Such a satellite will be located 35,919 kilometers above the Earth's surface (which is itself 6,378 kilometers from Earth's center).

If you feel uncomfortable with metric measurements, by the way, you can always convert kilometers to miles by dividing the number of kilometers by 1.609. You will then find out that a satellite in geosynchronous orbit is located at a mean distance of 22,324 miles above the Earth's surface.

† Those of you with nasty, suspicious minds will, I'm sure, check my calculations and catch me in arithmetical or conceptual errors.

If a satellite is in geosynchronous orbit, it might seem to you that it would move just in time with the Earth's rotation and it would therefore seem to remain in the sky in the same spot, day and night, for an indefinite period, if you're watching it (with a telescope, if necessary) from Earth's surface.

Not quite!

A satellite is in geosynchronous orbit at a mean distance of 42,298 kilometers from Earth's center, in whatever its plane of revolution might be. It might turn about the Earth from west to east (or east to west, for that matter), following a track above the equator. Or it might turn about the Earth from north to south (or from south to north) passing over both poles. Or it might be in any oblique orbit in between. All would be geosynchronous orbits.

If you were standing on Earth's surface, watching a satellite in geosynchronous orbit in a plane that formed an angle to Earth's equator, you would see its position change with reference to the zenith.

The satellite would mark out, in the course of a day, a figure eight, which astronomers call an "analemma." The greater the angle the orbit forms with the equator, the larger the analemma.

As an example, the Sun moves across the sky in an apparent orbit that is at an angle to the Earth's equator. For that reason, the position of the noonday Sun in the sky shifts from day to day. It marks out an analemma and, on a large globe, a proportionate analemma is usually placed in the empty spaces of the Pacific Ocean. From this analemma you can tell exactly how high in the sky the noonday Sun is on any day of the year (provided you adjust it to the latitude of the place where you are standing) and also exactly how many minutes the noonday Sun is short of zenith or past zenith on any day of the year. (It is *at* zenith on April 15 and August 30.)

This behavior of the Sun had to be allowed for in the old days of sundials, and "analemma" is, in fact, the Latin word for the block supporting a sundial.

A geosynchronous orbit need not be a perfect circle. It can be an ellipse of any eccentricity. It remains geosynchronous as long as the *mean* distance is correct. It can come in closer at one end of its orbit and retreat farther at the other end.

If, however, the orbit is elliptical as well as oblique, then the analemma is not symmetrical. One loop of the figure eight will be larger than the other. The more elliptical the orbit the greater the disparity in the size of the loops.

Thus, the Earth moves about the Sun in an ellipse of some slight eccentricity, so that the analemma formed by the apparent position of the noonday Sun from day to day in the course of the year is asymmetric. The northern loop is smaller than the southern, which is why the noonday Sun is at zenith about three weeks after the northern vernal equinox and three weeks before the northern autumnal equinox. If Earth's orbit were circular, the analemma would be symmetrical and the noonday Sun would be at zenith on the equinoxes.

Suppose, though, that a satellite is revolving about the Earth in Earth's equatorial plane. The orbit would form an angle of 0° with the equator and the analemma would be squashed to nothingness in the north-south direction.

If, however, the satellite were revolving in the equatorial plane in an ellipse, it would move faster than its mean speed in that part of its orbit where it was closer to the Earth than its mean distance, and slower when it was in the other portion. Part of the time it would outrace Earth's surface, and the rest of the time it would lag behind.

Viewed from the surface of the Earth, such a satellite would mark out a straight line, east and west, completing the back and forth motion in the course of a day. The more pronounced the eccentricity of the orbit, the longer the line.

But suppose that a satellite were not only revolving in Earth's equatorial plane, but were doing so in a perfect circle, west to east. In that case, the analemma would be totally degenerate. The motion north and south and the motion east and west would both disappear, and the satellite, when viewed from the Earth, would seem completely motionless. It would hang over one spot of the Earth indefinitely.

There is the difference between a geosynchronous orbit and a Clarke orbit. There are an infinite number of geosynchronous orbits, with any value of orbital inclination and orbital eccentricity. There is, however, only *one* Clarke orbit.

A Clarke orbit is a geosynchronous orbit with an orbital inclination of zero and an orbital eccentricity of zero. A Clarke orbit is exactly circular and exactly in the equatorial plane, and its value is precisely this: *Only* in a Clarke orbit will a satellite be motionless with respect to the Earth's surface.

This can be very useful. A satellite, motionless with respect to Earth's surface, will offer the simplest situation with respect to relaying communications, or to beaming energy. It was such an orbit Clarke visualized in his 1945 paper, hence "Clarke orbit."

Since there is only one Clarke orbit, and it is fairly close to the Earth, it represents a sharply limited resource. The length of the orbit is 265,766 kilometers—only 6.6 times the length of the circumference of the Earth (because the Clarke orbit is only 6.6 times as far from Earth's center as Earth's surface is).

Suppose you wanted to put a series of solar power stations in the Clarke orbit, and suppose it turns out that you can't really expect perfection. You can't place a satellite into the Clarke orbit *exactly*, and even if you could the gravitational perturbations of the Moon and the Sun would force it to jiggle about a bit. It might turn out, then, that safety would require placing the power stations at intervals of 1,000 kilometers. In that case, we could squeeze only 265 of them into the Clarke orbit, and that would present us with a limit to the amount of energy we could bleed from the Sun.

If there are other types of satellites that we would want in the Clarke orbit—communications satellites, navigational satellites, and so on—that would limit things even further.

One could imagine a particularly long satellite, with its long axis parallel to the Clarke orbit. Different types of functions could be placed along its length and these would never interfere with each other, for the satellite would move as a unit. The power stations at the two ends would not move relative to each other, or relative to the communications and navigational functions that might exist in between. In that way, a far greater number of working units could be squeezed into the Clarke orbit.

One might even picture a solid ring that filled the Clarke orbit, the sort of thing Larry Niven pictured in *Ringworld*. In that case, we could have functions of all sorts thickly strewn all along it. Such a ring is "metastable," however; that is, it would remain stably in orbit only as long as Earth remained in the exact center of the ring. If anything happened to nudge the ring slightly to one side (through gravitational perturbations, for instance), so that the Earth was no longer in the precise center of the ring, it would continue to drift in that same direction, would break up through tidal action, and parts would crash to Earth.

But then, there could be orbits related to the Clarke orbit that have values of their own.

Imagine a satellite in a circular orbit in the equatorial plane at a distance where its period of revolution is exactly *two* sidereal days, or three, or one and a half. A period of two sidereal days would mean that the satellite would move steadily, rising in the east and setting in the west, but from any point on the equator it would be seen directly overhead at forty-eight-hour intervals. Other periods that were simply related to the sidereal day would present their own patterns. (Even geosynchronous orbits that were inclined and eccentric and were therefore not Clarke orbits might be so arranged as to present simple patterns of behavior in the sky.)

I'm not sure what uses such patterns would have but they would be interesting from the standpoint of celestial mechanics. Let us refer to the whole family of orbits with inclination and eccentricity of zero as "Clarke orbits" regardless of distance and period of revolution. *The* Clarke orbit, where a satellite has a period of one sidereal day, would be a "Clarke-1 orbit." One in which the period was two sidereal days would be a "Clarke-2 orbit," and so on. We have the following distances from Earth's center then:

Orbit	Distance (kilometers)
Clarke-½	26,648
Clarke-1	42,298
Clarke-1½	55,410
Clarke-2	67,127
Clarke-3	87,980
Clarke-4	106,591
Clarke-5	123,679

The farther out such an orbit, the greater the effect of lunar perturbations upon it. I'm not enough of a celestial mechanic to be able to work out where a Clarke orbit would become large enough for perturbations to prevent its being useful for this purpose or that, but in time to come there will no doubt be computer simulations that will supply the answer—if such things don't already exist.

What works for Earth would work for any other astronomical body. Suppose, for instance, we wanted to place a satellite in orbit about Mars in such a way that it would seem to hover in one place in the sky as seen from the Martian surface. (Perhaps we would want continuous photographs of a particular spot on Mars over an extended period of time—as far as the inevitable interference of night, and of occasional dust storms, would permit.)

In the case of Mars, no geosynchronous orbit is possible, if we take our Greek seriously, since "geo-" applies only to Earth. You would have to speak of an "areosynchronous orbit." (I know, I know; people will speak of a geosynchronous orbit anyway, just as they casually say "lunar geology," when they really mean "selenology.")

Yet you can always speak of a Clarke orbit for any world. The term is not tied, etymologically, to the Earth. You can define a Clarke orbit as one in which an object will move about a larger object, with an orbital inclination and orbital eccentricity of zero, and with a period equal to the sidereal period of rotation of the larger object.

The question, then, is: What is the distance from the center of Mars to *its* Clarke orbit?

The Martian sidereal day is a bit longer than Earth's, since Mars rotates, relative to the stars, in 24.623 hours. This would have the effect of increasing the distance of the Clarke orbit, compared to that of the Earth, since the satellite need travel more slowly to keep up with the Martian rotation.

On the other hand, the intensity of the Martian gravitational field is only a tenth that of Earth's, so that the Clarke orbit would have to be closer to Mars if the satellite is going to be forced to circle Mars in a little over twenty-four hours. It is this second effect which is the greater, so that the Clarke orbit for Mars is at a distance of 20,383 kilometers from Mars' center.

Mars' Clarke orbit is just about half as far away from Mars as Earth's Clarke orbit is from Earth.

The farther Martian satellite, Deimos, is at a distance of 23,500 kilometers from Mars, and so is just outside the Clarke orbit. It therefore moves about Mars in slightly more than a Martian sidereal day, in 1.23 Martian sidereal days to be exact.

Any object lying outside the Clarke orbit (if we continue to assume all revolutions and rotations to be in the prograde direction, from west to east) will rise in the east and set in the west as viewed from the surface of the world it revolves about. This is true of Deimos, which rises in the Martian east and sets in the west, though it appears to move very slowly, since the Martian surface, as it rotates, nearly keeps up with it.

Mars' inner satellite, Phobos, has an orbit that lies *within* the Clarke orbit, since it is at a distance of only 9,350 kilometers from the center of Mars. It therefore revolves about Mars in less than a Martian sidereal day (0.31 such days, in fact) and overtakes the Martian surface.

Any object lying inside a Clarke orbit would appear to rise in the west and set in the east as viewed from the world about which the orbit exists—and that is indeed true of Phobos.

Jupiter is a particularly interesting case. It has an enormously intense gravitational field, one that is 318 times that of the Earth, and it also has a particularly fast rotation, making one complete turn on its axis in but 9.85 hours.

At what distance from Jupiter, then, would a satellite have to be in order to move about it in 9.85 hours? The answer is that Jupiter's Clarke orbit is at a distance of 158,500 kilometers from Jupiter's center. That is nearly four times the distance of Earth's Clarke orbit from Earth's center, despite the fact that a satellite moving about Jupiter must complete its circle in only two-fifths the time a satellite in Earth's Clarke orbit must if it is to maintain synchronicity.

Remember, though, that 158,500 kilometers represents the distance from Jupiter's *center*. Jupiter is a large, fat planet, however, and its equatorial surface is 71,450 kilometers from its center. This means that a satellite in a Clarke orbit about Jupiter would be only 87,050 kilometers above the visible surface of Jupiter's cloud layer.

Imagine, then, a satellite placed into a Clarke orbit about Jupiter in such a way that it is nearly above the Great Red Spot, which, alas, is not on the Jovian equator, so it wouldn't be *right* above it. What a continuous view it would have, during the five hours of daylight.

It could watch five hours on and five hours off for quite a long period of time, though there would be some complications. First, the Great Red Spot moves rather erratically and wouldn't stay in position indefinitely. Second, Jupiter's intense magnetic field might interfere with the satellite's workings. Third, we now know that Jupiter has a ring of debris close to its Clarke orbit and that might interfere, too.

Nevertheless, the sight would be a magnificent one if it could be managed, and since I have never heard of this being talked of (though that doesn't mean it hasn't been), I can at least dream that someday this *particular* Clarke orbit may be called the Asimov orbit.

Saturn, which, compared to Jupiter, has a slightly longer period of rotation (10.23 hours) and a considerably less intense gravitational field, has a Clarke orbit at a distance of 109,650 kilometers, or only 49,650 kilometers above Saturn's cloud layer.

There is one serious catch here, however. The gigantic ring system of Saturn lies in the planet's equatorial plane so that Saturn's Clarke orbit lies right within the rings, within Ring B, near the inner edge of the Cassini division.

This means that Ring B, the brightest portion of the ring system, lies almost entirely within the Clarke orbit, and therefore overtakes Saturn's surface as Saturn turns. If, from Saturn, the individual particles of Ring B (and of the rings lying still closer) could be made out, they would be seen to rise in the west and set in the east. Those particles lying beyond the Cassini division, however, would rise in the east and set in the west.

It might seem that we could pick some particle near the inner edge of the Cassini division and set up our instruments on it. We could choose one that was in a Clarke orbit. But then the myriad of particles that would be closer still to Saturn would block the visibility of that portion of Saturn's surface directly beneath.

There is a Clarke orbit about the Sun, too. It would be at a distance of about 26,200,000 kilometers from the Sun's center. This is less than half the distance of Mercury from the Sun.

Back in the late nineteenth century, there was considerable speculation that a small planet, called Vulcan, existed inside Mercury's orbit (see "The Planet That Wasn't," in the book of the same name, Doubleday, 1976).

Unfortunately, Vulcan does not exist. What a pity! Its orbit would have been bound to be close to the Sun's Clarke orbit. Suppose it were exactly in the Clarke orbit, and that we could reach it and place our instruments upon it, and that those instruments could withstand the fiery furnace of the nearby Sun.

Imagine the view of sunspots beneath. They could be followed at close range for much of their lifetime. (There would be a complication in that the Sun's surface, at different latitudes, rotates at different speeds, so that the sunspots would gradually seem to drift away.)

Venus has a very slow period of rotation (243.09 days) and the intensity of its gravitational field in only 0.815 times that of the Earth. You would expect a distant Clarke orbit in that case, and you would get it. Venus' Clarke orbit lies at a distance of 1,537,500 kilometers from the planet's center, just four times as far from Venus as the Moon is from Earth. It would be a pretty useless Clarke orbit at that distance.

Mercury's Clarke orbit would be 240,000 kilometers from Mercury, or considerably less than the distance of the Moon from the Earth.

And that's about all the publicity I'm going to give old whatzis-name.

XII

Ready and Waiting

I have just returned from an "Astronomy Island" cruise to Bermuda. The idea is to visit a site on that beautiful island where we can gaze at various objects in its clear sky through a variety of telescopes set up by one or another of the enthusiasts who have come.

It is always the sky of July or August, with the week carefully chosen for the absence of the Moon. Scorpio is always prominent in the southern sky, winding its S-shaped way down toward the horizon.

Immediately below and to its left (from our vantage point) are eight stars that mark out a perfect teakettle to my eyes, and that is Sagittarius. At the star marking the spout of the teakettle, the Milky Way curls up and, to the left, is like faint steam.

That site in Sagittarius is the brightest part of the Milky Way, and if you stare in that direction, you are staring toward the center of the Galaxy.

It's rather exciting to know that even though you can't see through the dust clouds, somewhere out there—right in the direction your eyes are gazing—there is a region of unimaginable turbulence that includes, in all likelihood, an enormously massive black hole.

And yet, ever and anon, my eyes would turn to Antares, the brightest star in the constellation Scorpio, and I would watch it fixedly for a while.

Maybe— Maybe— Maybe—

The chances were one in a large number of trillions that anything would happen to it while I watched, but, just in case, I wanted to be ready and waiting.

But, of course, nothing ever happened.

What is it I expect? Well, let's begin at the beginning.

About 130 B.C., the Greek astronomer Hipparchus (190–120 B.C.) prepared the first star catalog. He listed nearly 850 stars, using the names they were then given, and gave their latitude and longitude with respect to the ecliptic (the path followed by the Sun against the starry background) and the Sun's particular position at the vernal equinox.

Why did he do it? According to the Roman author Pliny (23–79), writing two centuries later, it was because he had "discovered a new star."

Mind you, before the invention of the telescope, it was taken for granted by almost all stargazers that the stars were all visible to people with acute vision. The notion of an invisible star seemed like a contradiction in terms. If it was invisible, it wasn't a star.

Yet stars vary in brightness and most of them are so dim they are difficult to see. Might it not be possible that some—a few, at least— were so dim that they could not be made out at all by human eyesight, however acute? To us, today, thinking about it with the brilliance of hindsight, the possibility seems so overwhelmingly logical that we wonder how anyone could fail to see it.

The trouble is that, until about four and a half centuries ago, human beings lived in a homocentric Universe, and firmly believed that everything in the Universe had been created only in order to exert some effect on human beings. (Most human beings live in such a Universe even today.)

People might argue that the stars existed only because they were so beautiful that they pleased our eyes and stirred us to wonder and romance.

Or, to be more utilitarian, they might argue that the stars formed a complex cryptogram, against which movable objects, such as the Sun, the Moon, planets, comets, and meteors, marked out paths from which hints for human guidance could be obtained.

Or, to be more lofty, they might argue the stars were intended to stir the soul to a sense of its own unworthiness and to give hints of the existence of a transcendent entity beyond human grasp or understanding. ("The heavens declare the glory of God; and the firmament showeth his handiwork." Psalms 19:1.)

In a homocentric Universe, it simply makes no sense to imagine an invisible star. What would be its purpose? Being unseen, it couldn't serve either aestheticism, utilitarianism, or religion.

Yet Hipparchus, having gazed at the heavens sufficiently and having spent enough time plotting the position of the planets against the starry background to know the pattern of the thousand brightest stars by heart, looked at the night sky and saw a star that hadn't been there the last time he had looked.

He could only assume it was a *new* star, one that was freshly formed. And only temporarily, too, for eventually it vanished again. (Pliny doesn't say so, but we can be sure it did.)

It must have seemed to Hipparchus that such a heavenly intrusion was a notable event, and he must have wondered if it happened frequently. To be sure, there had been no prior reports of new stars, but such a silent insertion might simply have gone unnoticed. Few knew the heavens as Hipparchus did and a slight irregularity would pass unheeded. So he prepared his catalog, in order that some future stargazer, at the merest suspicion of novelty, might consult it to see if a star was actually supposed to exist at the position one had been sighted.

Occasionally, though infrequently, new stars were noted in the centuries after Hipparchus. A particularly noticeable one appeared in the constellation Cassiopeia on November 11, 1572. A twenty-six-year-old Danish astronomer, Tycho Brahe (1546–1601), observed it carefully, and wrote a fifty-two-page book concerning it which made him, at once, the most famous astronomer in Europe.

Tycho (he is usually known by his first name) gave the book a long title that is usually boiled down to *Concerning the New Star*. Since he wrote in Latin, however, the title should really be *De Nova Stella*. From that time to this, a "new star" has been called a *nova*, which is the Latin word for "new."*

*The Latin plural is *novae*, but a steadily lessening interest in Latinic minutiae has led to "novas" as the usual plural. —We also say "formulas" instead of "formulae," and any day I expect to hear people speaking of "two memorandums."

And then, in 1609, Galileo (1564–1642) constructed his first telescope, turned it on the heavens, and noted at once that it brightened
each star in appearance, and that many stars, too dim to see ordinarily,
were brightened into visibility by it. There were, he found, numerous
invisible stars in existence, greater in number than the visible stars
were. If any one of them happened to brighten sufficiently, for any
reason, it would become visible to the unaided eye and, in pretele-
scopic days, would appear to be a "new" star.

In 1596, for instance, the German astronomer David Fabricius
(1564–1617) had noted a third-magnitude star, in the constellation of
Cetus, which faded and eventually disappeared. He considered it another temporary star that had come and gone, as those of Hipparchus
and Tycho had. In the course of the next century, however, the star
was seen in the same place on several occasions. With the use of the
telescope, it was found to be there at all times, but to vary irregularly
in brightness. At its dimmest, it could not be seen by the unaided eye,
but it could brighten to different degrees of visible brilliance and, in
1779, it even temporarily reached the first-magnitude mark. It came to
be called "Mira" ("wonderful"), though its more systematic name is
Omicron Ceti.

Nowadays, any star is classified as a nova if it brightens sharply and
suddenly, though it may be so dim to begin with that even at its
brightest it remains invisible to the unaided eye. Stars may also brighten
and dim *regularly*, but they are then "variable stars" and are not
considered novas. On the other hand, novas are usually classified as a
variety of variable star.

Now that we have the telescope to help us, novas are not the marvel
and rarity they once were. About 25 per year occur in our Galaxy on
the average, though most are hidden from us since dust clouds keep
us from seeing all but our own corner of the Galaxy.

Generally, a nova comes without warning and is detected only after
it has suddenly brightened. I don't think anyone has ever happened to
be watching a star and to have actually caught it begin to brighten.
Once it has brightened and been detected, however, it can be observed
after it has faded out to what, presumably, it was before.

More and more such "post-novas" were studied and, by the 1950s,
it was clear that every one of them, without exception, was a close
binary. Novas were found to be a pair of stars circling a common

center of gravity, and so close as to involve considerable tidal influence. One of the pair was always a white dwarf, the other a normal star.

What happened was plain. The tidal influence of the white dwarf on its ordinary partner pulled hydrogen-rich matter out of the latter. This matter would form a ring about the white dwarf, and the matter would slowly spiral in toward it. As the matter approached the white dwarf, it would be subjected to an intensifying gravitational pull that would condense it and produce hydrogen fusion within it. The white dwarf would always shine a bit brighter than it would if unaccompanied because of the sparkling of the hydrogen cloud stolen from its companion star.

Every once in a while, however, unusually large gouts of matter would leak over from the main-sequence star (because of unusual activity on its surface, no doubt) and a relatively huge amount of hydrogen would descend upon the white dwarf. The resulting explosion would supply many times the light that the white dwarf itself could deliver and, as seen from Earth, the star (showing in our eyes as a single dot of light, including both companions) would suddenly become much brighter than it had been. And then, of course, the hydrogen supply would eventually be consumed and the star would fade back to what it had been before—until the next large delivery.

That's not all there is to the story, though.

In 1885, a star was seen in the central regions of what was then known as the Andromeda Nebula, a place where no star had been visible before. It lingered on for a period of time, then slowly faded and disappeared. At its peak, it was not quite bright enough to be seen by the unaided eye, and it was considered a rather poor specimen. That it was bright enough to deliver nearly as much light as the entire Andromeda Nebula was not considered important.

But suppose the Andromeda Nebula was not a relatively nearby collection of dust and gas (as most astronomers then supposed) but was a very distant collection of stars, one that was as large and as complex as our own Galaxy. Some astronomers suspected *that*.

In the 1910s, an American astronomer, Heber Doust Curtis (1872–1942), studied the Andromeda Nebula and began to observe tiny brightenings that took place within it. These he attributed to novas. If the Andromeda Nebula were very distant, the stars within it would be so exceedingly faint as seen from Earth that the nebula would appear as a mere fog. The novas would brighten to the point where they could be made out individually in a good telescope but would still be exceedingly faint as compared to the stars of our own Galaxy.

Curtis spotted large numbers of novas in the Andromeda Nebula, dozens of times as many as would appear in the same time in other similarly sized patches of sky. His conclusion was that the nebula was indeed a galaxy and contained so many stars that the novas among them would be numerous. He turned out to be right. The Andromeda Galaxy (as it is now known) is about 700,000 parsecs away, over thirty times as far away from us as the farthest star in our own Galaxy. (One parsec, by the way, is equal to 3.26 light-years.)

In that case, how could the nova of 1885 have nearly achieved unaided-eye brightness? In 1918, Curtis suggested that the nova of 1885 was an exceptional, extraordinarily bright nova. In fact, if the Andromeda Nebula was really a galaxy as large as our own, then the nova of 1885 was shining with all the brilliance of an entire galaxy, and was many billions of times as luminous (temporarily) as our Sun. Ordinary novas are only a few hundred thousand times as luminous (temporarily) as our Sun.

In the 1930s, the Swiss astronomer Fritz Zwicky (1898–1974) made a painstaking search for other-galaxy novas that blazed up to galactic brilliance and he called this extrabright breed "supernovas." (The nova of 1885 is now called "S Andromedae.")

Whereas a nova can repeat many times, doing so each time it gets a large supply of hydrogen from its companion star, the supernovas are one-shots.

A supernova is a large star that has consumed all the fuel at its core and that can no longer maintain itself against the pull of its own gravity. It has no alternative, then, but to collapse. As it does so, the kinetic energy of inward motion is converted to heat, and the hydrogen that still exists in its outer regions is heated and compressed to the point where fusion reactions are ignited. All the hydrogen goes more or less at once and the star explodes and, giving off all its energy supply in a very short time, temporarily brightens to a glow that rivals that of a whole galaxy of ordinary stars.

What is left of the star after the explosion collapses to a tiny neutron star and, of course, never explodes again.

Supernovas are much rarer than ordinary novas, as you might well suspect. At most, there would be one supernova for every 250 or so ordinary novas. In a galaxy the size of ours, there might be one every ten years, but most of them would be hidden by dust clouds lying between the explosion and ourselves. Perhaps once every three centuries or so, a supernova would take place in the relatively small corner of our Galaxy that is visible to our eyes and our optical telescopes.

Naturally, supernovas are much more spectacular than novas when the two are seen at comparable distances. The question, then, is this: Has a supernova ever been viewed in our own corner of the Galaxy?

The answer is: Yes!

The "new star" viewed by Tycho was undoubtedly a supernova. It brightened rapidly until it was brighter than Venus! It was visible in the daytime and by night cast a faint shadow. It stayed very bright for a couple of weeks and remained visible to the naked eye for a year and a half before fading altogether out of sight.

In 1604, another supernova flared and was observed by the German astronomer Johannes Kepler (1571–1630). It was not as bright as Tycho's supernova, for it never grew brighter than the planet Mars. But then Kepler's supernova was farther away than Tycho's had been.

This meant that two supernovas blazed brightly down on Earth within a space of thirty-two years. If Tycho, who died in 1601, at the age of fifty-four, had lived but three more years, he would have seen both of them.

And yet (such is the irony of events) in the nearly four hundred years since then, *not one local supernova has showed up*. Astronomers' tools have advanced unbelievably—telescopes, spectroscopes, cameras, radio telescopes, satellites—but no supernovas. The closest one visible since 1604 was S Andromedae.

Were there any supernovas *before* Tycho's?

Yes, indeed. In 1054 (possibly on July 4, in a remarkable celebration in advance), a supernova blazed forth in the constellation of Taurus and was recorded by Chinese astronomers. It, too, was brighter than Venus at the start, and it, too, faded only slowly. It remained visible to the eye in daytime for three weeks, and at night for two years.

Except for the Sun and Moon, it was the brightest object in the sky that appeared in historic times. Oddly enough, no observations of the

Taurus supernova have survived from any European or Arabic source.

There is a follow-up to this story, though. In 1731, an English astronomer, John Bevis (1693–1771), first observed a small patch of nebulosity in Taurus. The French astronomer Charles Messier (1730–1817) published a catalog of foggy objects forty years later, and the Taurus nebulosity was first on the list. It is sometimes known as M1, therefore.

In 1844, the Irish astronomer William Parsons (Lord Rosse) (1800–67) studied it, and noting the clawlike processes extending in all directions, called it the Crab Nebula. That is the generally accepted name today.

Not only is the Crab Nebula at the precise spot recorded for the 1054 supernova, but it is clearly the result of an explosion. The gas clouds within it are being driven outward at a speed which can be measured. When calculated backward, it is apparent that the explosion took place nine centuries ago.

A tiny star was detected at the center of the Crab Nebula in 1942 by the German-American astronomer Walter Baade (1893–1960). In 1969, that star was recognized as a "pulsar," a rapidly rotating neutron star. It is the youngest pulsar known, rotating thirty times a second, and it is all that is left of the giant star that exploded in 1054.

The Crab Nebula is about 2,000 parsecs away, not very far as distances go in the Galaxy, so it is not surprising that the display was so magnificent. (The more distant supernovas of 1572 and 1604 have left no clearly recognizable remnants.)

There may, however, have been an even more astonishing event in prehistoric times.

About 11,000 years ago, at a time when, in the Middle East, human beings were soon to develop agriculture, a star that was only about 460 parsecs away (less than a quarter the distance of the 1054 supernova) exploded.

At its peak, the supernova may have approached the full Moon in brightness, and this appearance of a second moon that did not move against the starry background of the sky, that did not show a visible disk or phases, that slowly faded but did not disappear for perhaps three years must have totally astonished our not-yet-civilized ancestors.

Naturally, no records exist from that time (though there are some symbols on prehistoric sites that may indicate that something unusual had been noted in the sky) but we have indirect evidence.

In 1930, the Russian-American astronomer Otto Struve (1897–1963) detected a large area of nebulosity in the sky in the constellation of Vela, which is far down in the southern sky, and is totally invisible from positions as far north as New York.

This nebulosity is in the form of a shell of gas and dust that was driven out from the Vela supernova 11,000 years ago. It is much the same sort of phenomenon as the Crab Nebula, but it has been expanding for over twelve times as long a period of time so that it is much larger.

It was investigated in detail in the 1950s by an Australian astronomer, Colin S. Gum, and is known as the Gum Nebula in consequence. The near edge of the nebula is only about 92 parsecs from us and, at the rate at which it is now expanding, it may move across the Solar System in about 4,000 years or so. However, the matter it contains is so thin by now (and will be thinner in 4,000 years) that it is not likely to affect us in any perceptible way.

When will the next visible supernova appear? And what star is it that will explode?

If we could only have observed a nearby supernova in the process of explosion with the full battery of modern instruments, we might be able to answer the question with considerable precision, but, as I said, we are completing a four-century desert as far as these events are concerned.

Nevertheless, we know a few things. We know, for instance, that the more massive a star is, the more quickly it consumes its core fuel, the shorter its life as an ordinary "main-sequence star," and the more rapid and catastrophic its collapse is.

Even a star as large as our Sun will blow off only a minor fraction of its mass, when the time comes, and will then collapse sedately into a white dwarf. The blown-off mass will expand outward, forming what is called a "planetary nebula," because it is seen as a ring about a star and such a ring was thought, a hundred years ago, to serve as precursor to planet formation.

In order to have a real explosion and a collapse to a neutron star, the mass of the star has to be 1.4 times the mass of the Sun as an absolute minimum and, very likely, a good explosion will require a star that is up to ten or twenty times the mass of the Sun.

Such stars are rare indeed. There may not be more than one star in 200,000 or so that is massive enough for a good supernova. Still, that leaves about 100,000,000 of them in our Galaxy, and perhaps 300,000 in our visible corner of it. These giant stars have a lifetime on the main sequence of only 1 to 10 million years (as compared to about 10 to 12 billion years for the Sun) so, on the astronomic scale, they explode frequently.

You might wonder why the giant stars haven't exploded by now if supernovas are forming once every decade. At that rate, all the giant stars would be gone in 1 billion years and the Galaxy is nearly 15 billion years old. In fact, if they endure only a few million years before exploding, why did they not all disappear in the childhood of the Galaxy?

The answer is that more are constantly being formed and that all the giant stars that exist anywhere in the Galaxy now came into being only 10 million years ago or less.

There's no way in which we can watch all of them constantly, but there's no need to. The beginning of the slide to supernova-dom is easily noticeable, and we need concentrate only on those that have made that beginning.

When a star reaches the end of its stay on the main sequence, it begins to expand. It reddens as it does so, since its surface cools with expansion. It becomes a red giant. This is a universal step. Some time in the future—anywhere from five to seven billion years from now— our Sun will become a red giant, and the Earth may be physically destroyed in the process.

The more massive a star, the larger the red giant stage, of course, so what we must look for are not just massive stars, but massive red giants.

The nearest red giant is Scheat, in the constellation of Pegasus. It is only about 50 parsecs away and its diameter is about 110 times that of the Sun. This is small as red giants go, and if this is as big as it is going to get, it is probably no more massive than the Sun and will not ever be a supernova. If it is still expanding, it has a considerable way to go before exploding.

Mira, which I mentioned earlier in the chapter, is 70 parsecs away, has a diameter 420 times that of the Sun, and is definitely more massive than the Sun.

There are three red giants more massive still, however, each of them being about 150 parsecs away. One of these is Ras Algethi, in Hercules, with a diameter 500 times that of the Sun, and another is Antares, in Scorpio, with a diameter 640 times that of the Sun. (That is why I can't help but keep an eye on Antares when I am on Bermuda. Just imagine if I happen to be looking at it at the very moment it decides to blow, and I watch it increase in brightness to well beyond that of Venus in the space of an hour or less. Wow!)

Larger still is Betelgeuse, in Orion. It is not only large, but it is pulsating, and its brightness varies. This might indicate the kind of instability that could well precede explosion. It is as though the star keeps collapsing and then, as pressure rises at its core, a little more energy is squeezed out, so that it expands again. (Such pulsation is also found in Mira.)

Astronomers, however, have now discovered what may be the best candidate. It is Eta Carinae, in the constellation Carina. This is an enormous red giant, even larger than Betelgeuse, and it has a mass estimated to be a hundred times that of the Sun.

It is surrounded by a cloud of dense and expanding gas, which it may be giving off in what we might consider its death throes. What's more, it shows marked and irregular changes in brightness, either because it is pulsating, or because we sometimes see it through breaks in the surrounding cloud and sometimes see it obscured.

It can become bright indeed. In 1840, it was the second brightest star in the sky, surpassed only by Sirius (although, to be sure, Eta Carinae is well over a thousand times as far away as Sirius is).

Right now, Eta Carinae is too dim to see with the unaided eye. However, its radiation is absorbed by the cloud about it, and reradiated as infrared. The energy it is emitting can be grasped when it is realized that it is the strongest infrared-radiating object in the sky outside our own Solar System.

Finally, astronomers have recently detected nitrogen in the cloud it is ejecting and they judge that this, too, indicates a late stage in the pre-supernova changes. The betting seems to be that Eta Carinae can't last more than 10,000 years at most. It might also blow up tomorrow. Since it takes 9,000 years for light to travel from Eta Carinae to us, it is possible that the star has already exploded and that the light of that explosion is well on its way to us. —In any case, astronomers are ready and waiting.

The catches? Two!

First, Eta Carinae is about 2,750 parsecs away, nearly twenty times as far away as Betelgeuse, and the brilliance of the supernova will be somewhat dimmed by the extra distance.

Second, the constellation, Carina, is far in the southern sky, and the supernova, when it comes, will not be visible in Europe or in most of the United States.

But then, you can't have everything.

XIII

Dead Center

I received a letter today from someone who, knowing I lived in New York City, wondered how anyone could possibly bear to live in a large city, or *any* city. He himself (he said) lived in a town of 5,000 people and was planning to move to one of 600 people.

You can well imagine how indignant I grew at this.

My first impulse was to reply and tell him, haughtily, that the only advantage of living in a small town was that it robbed death of its terrors. —But I fought that down and didn't reply at all.

To each his own!

And yet it seems to me that there must be something in each of us that feels a certain yearning for "centeredness." A large city is the center of a region. Beyond it are the "outskirts," the "suburbs," the "hinterland." The words themselves indicate that the city is the essence, while everything else is subsidiary.

I get a certain pleasure in knowing that I live not merely in a city but in Manhattan, the center of New York City, a region so unique in many ways that I honestly believe that Earth is divided into two halves: Manhattan and non-Manhattan.

I even boast that I live at the very geographical center of Manhattan, though that is not exactly true. The actual central point is in the aptly named Central Park, and, as nearly as I can determine, I live about half a kilometer west of that point.

Nor am I alone in this "centrocentric" attitude. Everyone is. Statisticians go to a lot of trouble to figure out the exact geographical center of the United States. (If you're interested, the geographical center of the forty-eight contiguous states is in Smith County, Kansas, near the town of Lebanon. If you add Alaska and Hawaii, the center moves northwestward to Butte County, South Dakota, west of the town of Castle Rock.)

You could easily find the center for any region, nation, continent, or ocean. I suppose that anyone in the world can carefully select an area in such a way as to place himself in the center of something. (The county seat of Smith County, Kansas, is in the geographical center of that county and proudly calls itself Smith Center.)

That minimizes the pleasure of centrocentrism, however. If everyone can be at the center of something, where's the value?

We have to stop fooling around and figure out some way of deciding the center of the Earth itself, something unique to the entire world.

In the days when people thought Earth was a flat disk with the sky coming around on all sides to meet it at the horizon, it must have seemed to each person that he himself was at the very center of the world. It didn't take much of an advance in sophistication, however, to come to the realization that there was more to Earth than that which was visible within the circular horizon. The "egocentric Universe" had to be dismissed.

Nevertheless, there was resistance to thinking the center to be very far from one's own feet. If one was not the center, then one's culture ought to be—in particular, the most highly regarded spot related to that culture, if there was one. Thus, the ancient Jews were quite certain that Jerusalem was at the center of the Earth, and placed the precise central point at the Holy of Holies within the Temple at Jerusalem.

The Greeks, for very similar reasons, felt that Delphi was at the center of the Earth, and placed the precise central point at the cleft over which the Pythian priestess sat, inhaled the fumes, and made the incoherent sounds that were translated into prophecies.

And (not entirely jokingly), the old Yankee Brahmins thought Boston was at "the hub of the Universe" and placed that hub precisely at the State House.

I suppose every group devises a "culturocentric Universe," either literally or symbolically.

The fun was ruined when it turned out that the Earth was not flat, but spherical (not *exactly* spherical, but let's not quibble). The surface of a sphere has no center.

To be sure, a *rotating* sphere has two special points on its surface, the North Pole and the South Pole, but each exists in such an undesirable location that they lose their value. No one would feel any special pride in living at either pole; nor would anyone be moved to establish a central religious shrine at either.

Quite arbitrarily, we divide the Earth's surface into degrees of latitude and longitude, and there is a unique place that is at 0° latitude and 0° longitude. That is the result of human convention, however, and is located in the Gulf of Guinea about 625 kilometers due south of Accra, the capital of Ghana. Who is going to establish a religious shrine in the ocean?

There are other arithmetical coincidences that might be drawn upon. For instance, a mere 130 kilometers due west of the Great Pyramid is a point which is at 30° north latitude and 30° east longitude. There are people who have seriously suggested that the ancient Egyptians had some mystic purpose in building their pyramids near the "double-thirty." (Of course, it wasn't double-thirty till some 4,200 years after the building of the Pyramids, when the British set up the prime meridian so as to make it run through the Greenwich Observatory near London for decidedly culturocentric reasons of their own. The double-thirty Pyramid connection boils down, as so many things do, to coincidence, therefore, and it would take near-madness to argue anything else.)

What it comes down to is that, when dealing with a sphere, we must abandon the surface altogether, if we want to be sensibly centric. We must deal with the true center, the dead center, which is equidistant from every point on the surface. The center of Earth is 6,378 kilometers straight down, no matter where you are standing (provided you consider Earth a perfect sphere and ignore the equatorial bulge and the minor unevennesses of hill and dale).

None of us has the privilege of living at the center of the Earth (or wants it), but then, none of us is closer to the center or farther from it than anyone else by any significant amount, and that is just as good. If we are "eccentric" (in the literal sense), we are all equally eccentric.

The ancient Greek philosophers were the first who had to deal with a spherical Earth and they continued to do their best to make the Universe as egocentric as possible. (I don't blame them, you understand. I would surely have done the same.)

They made the center of the Earth the center of the Universe as a whole. They eventually visualized the Earth as surrounded by a series of concentric spheres which held, in turn, Moon, Mercury, Venus, Sun, Mars, Jupiter, Saturn, and the stars, in that order, moving outward. The center of each of these spheres coincided with that of Earth.

The mathematics that had to be used to predict the position of the planets in the sky, against the background of the stars, on the assumption of such a "geocentric Universe," was worked out by Hipparchus of Rhodes about 130 B.C. and perfected by Claudius Ptolemaeus (100–70) about A.D. 150.

There were some Greek astronomers, notably Aristarchus of Samos (310–230 B.C.) and Seleucus of Seleucia (190–120 B.C.), who disagreed, but they were ignored.

It was not till 1543 that the Polish astronomer Nicolas Copernicus (1473–1543) was able to show that the mathematics used to predict planetary positions could be simplified, if one assumed that the Sun were at the center of the Universe in place of Earth. This would make it a "heliocentric Universe."

Copernicus thought the Sun was surrounded by concentric spheres which held Mercury, Venus, Earth (and its attendant Moon), Mars, Jupiter, Saturn, and the stars, in that order, moving outward. The center of each of these spheres coincided with that of the Sun.

This was not just a matter of placing particular individuals off center as in the case of a culturocentric Universe, or even all people off center as in the case of a geocentric Universe. The vast Earth itself was placed off center and it therefore took fifty years for astronomers generally to accept the heliocentric Universe. (If we put it to a vote of the Earth's population generally, even today, I think heliocentrism would still lose out.)

In 1609, the German astronomer Johannes Kepler did away with the spheres altogether. He showed that the actual movement of the planets across the sky could be better explained by supposing them to be moving in elliptical orbits—and this view of the Solar System has been retained ever since, with only the most minor of refinements.

Ellipses have centers, as circles and spheres do, but the center of the ellipses that mark the planetary orbits do not coincide with the center of the Sun. Rather, the Sun is at the focus of each ellipse, that focus being to one side of the center.

In 1687, the English scientist Isaac Newton (1642–1727) presented his Law of Universal Gravitation, and from that it could be seen that the Solar System, as a whole, had a center of gravity. This center of gravity might be considered as motionless, while all the bodies of the Solar System (including the Sun!) revolved about that center in a fairly complicated fashion. The Sun was much nearer to the center of gravity, at all times, than was any other body of the Solar System so that, as a rough approximation, one could still say that all the planets revolved about the Sun.

The center of gravity was often so far from the Sun's center (more or less in the direction of Jupiter) as to be beyond its surface, but, on the scale of the Solar System, to have the center of gravity of the system as much as 1,000,000 kilometers from the center of the Sun means little, and we can still view the Sun as approximately the center of the system.

Nevertheless, it is the center of gravity of the Solar System that is at the center of the Universe in the Copernican sense, and we ought to speak of a "systemocentric Universe" rather than a heliocentric one.

It was fair enough to speak of a systemocentric Universe even in Newton's day, since (for all anyone knew) the stars might be evenly distributed around the Solar System and might all be affixed to a thin solid shell (or "firmament") just beyond the farthest planet. That certainly fit the appearance (and perhaps a majority of the Earth's population still believe this).

Nor did the discovery of the true distance of the planets, and of new members of the Solar System—new planets, new satellites, myriads of asteroids and comets—in itself affect the systemocentric view. That had to hold as long as the firmament existed, no matter how far off it might be, or how diverse or numerous the objects within it.

The first blow to the firmament came in 1718, when the English astronomer Edmund Halley noted that at least three bright stars, Sirius, Procyon, and Arcturus, had changed their positions markedly since Greek times. Other astronomers detected such changes in position for other stars.

It became clear that the stars were not fixed to the firmament, after all, but crawled along it at various speeds and in varying directions— which made it doubtful that the firmament existed at all. It became possible (indeed, almost irresistible) to suppose that the stars occupied a volume within which they moved randomly, like bees in a swarm. If all moved at about equal speeds, those closest to the Solar System would seem to move most rapidly, while those farthest would seem to move so slowly that the motion would not be apparent even over extended periods of time.

In 1838, the German astronomer Friedrich Wilhelm Bessel (1784–1846) worked out the distance to a star for the first time. The distances to other stars were quickly determined. It turned out that even the nearest star is 1.3 parsecs away. The distance from the Sun to the nearest star is 9,000 times as great as the distance from the Sun to the farthest large planet. Other stars are much farther away still; clearly hundreds, perhaps, thousands of parsecs away.

Nevertheless, if the stars were finite in number and were distributed with spherical symmetry about the Sun (however great their distances might be) the Universe might still be systemocentric.

Consider—

All the objects in the Solar System, including the Sun, revolve about the center of gravity of the Solar System. (Some objects, the satellites, do so while revolving about the center of gravity of a particular satellite system. Thus, the Moon and Earth revolve about the center of gravity of the Earth-Moon system, and each is carried along as that center of gravity revolves about the overall center of gravity of the Solar System.) The objects in the Solar System need not be revolving all in the same plane. To be sure, the planets very nearly are, but if you include the asteroids and comets, the revolving objects form a thick spherical shell about the center of gravity of the Solar System, with the Sun very near that center.

In the same way you can imagine all the stars (each perhaps with an attendant system of planets) revolving about the center of gravity of the entire star system, and if that center of gravity coincided, or nearly coincided, with the center of gravity of the Solar System, then the entire Universe was still systemocentric.

Of course, the larger the Universe proved to be and the more certain it was that it consisted of millions of stars each rivaling the Sun in size, the less reason there seemed for that Universe to be systemocentric. Why should the entire vast Universe, all those millions of stars, have *us* as their center, and why should all revolve about *us*?

To the religious, there would be no mystery. It was the way God designed the Universe. Indeed, from the fact that the Universe was systemocentric, one could deduce that the Solar System was of peculiar importance, and that could only be so because human beings exist there, and that this is so could only be because we are formed in the image of God. In this way the systemocentric nature of the Universe becomes a magnificent "proof" of the existence of God.

To the nonreligious, the only possible response to the situation is that that is the way it seems to be and that, perhaps, at some time in the future, as our knowledge increases, we will understand the matter better.

The discomfort of systemocentrism could be removed only if there were some reason to think that it didn't exist or that, if it did, it was merely circumstance, and not part of the intrinsic design of the Universe.

For instance, suppose the Universe were infinite in size and that the stars, in every direction, stretched out forever and forever. (The German scholar Nicholas of Cusa (1401–64) had maintained exactly this as early as 1440.)

In that case, there would be no center. Every point within an infinite sphere has just as much right to consider itself a center as any other, and there is no privileged position at all. (The situation is precisely that of the surface of a sphere on which there is no center and no privileged position.)

If the Universe were infinite, in short, it would *seem* that we were at the center, but that would be true no matter in what planetary system we were located. (The maintenance of systemocentricity would then be as naïve as an individual's view that he was the center of the Universe because he was at the center of the circle of the horizon.)

In 1826, however, the German astronomer Heinrich Wilhelm Matthäus Olbers (1758–1840) pointed out that if the Universe were infinite in size and contained an infinite number of stars evenly spread out in all directions, the entire sky would be as bright as the circle of the Sun. There are a number of ways in which one might explain the blackness of the sky in view of this (see "The Black of Night," in *Of Time and Space and Other Things*, Doubleday, 1965) but the simplest is to take such blackness as evidence of the fact that the Universe is *not* infinite in size, and the stars are *not* infinite in number. In that case, the Universe, by nineteenth-century thinking, *must* have a center, and the Solar System seemed to be there.

By that time, though, William Herschel had made a particularly interesting discovery.

By 1805, he had spent more than twenty years determining the proper motion of various stars (their motions, that is, relative to very dim and, therefore, supposedly very distant stars—stars too distant to show any motion at all). As a result, he was able to demonstrate that in one part of the sky the stars, generally, seemed to be moving outward from a particular center (the "apex"). They didn't do so uniformly, or universally; but they did so, on the whole.

In a place in the sky directly opposite to the apex, the stars generally seemed to be moving inward toward an imaginary center (the "antiapex"). The apex and antiapex were just about 180 degrees apart.

One way of explaining this was to suppose that what Herschel had detected was what was actually happening: that the stars were moving apart in one part of the sky and coming together in the opposite part, moving around the stationary Solar System as they did so and giving it a wide berth. If that were so, what a testimony it would be to the special position of the Solar System.

An alternate interpretation of the observation is, however, possible. It is that the Sun itself is moving relative to the nearby stars (those sufficiently nearby to show detectable proper motion).

Suppose yourself, for instance, to be in the midst of a forest of randomly placed trees, each quite distant from its neighbors. As you looked about in any direction, the trees nearby would be spread far apart, but those that were far away would seem closer together. If you moved in a certain direction, the trees in that direction would be closer and closer to you as you moved and would seem to spread farther and farther apart. In the opposite direction, you would be moving away from the nearest trees and they would seem to be coming closer together.

This is a common effect of perspective, so common we scarcely notice it, unless we are little children. Our minds make allowance for it and we are never deluded into thinking that the trees are actually either separating or coming together.

Once you think of that it makes much more sense to suppose that the "Herschel effect" is indeed the result of the Sun moving. No astronomer supposes any other explanation is necessary. Thanks to the observations made since Herschel's time, astronomers are now quite certain that the Sun is moving (relative to the nearer stars) in a direction toward a point in the constellation of Lyra at a speed of 20 kilometers per second.

How does this affect the systemocentricity of the Universe?

If the Sun is moving, undoubtedly carrying its planetary system (including Earth) with it, then it clearly cannot be the motionless center of the Universe.

There must, however, still be a motionless center of gravity of the star system about which all the individual stars are revolving, and if the Solar System is not at that point, it nevertheless seems to be near it.

Just as the Sun moves in a tight orbit about the center of gravity of the Solar System, the Solar System may move in a tight orbit about the center of gravity of the star system. In that case, if the Universe is not systemocentric, it is nearly systemocentric.

On the other hand, it may be that the Solar System moves in a very elongated orbit about the center of gravity of the star system (like a comet moving about the center of gravity of the Solar System). In that case, for much of its history, the Solar System would be very far from the center of gravity, but *right now* it happens to be close to it. Considering the size of the Universe and the rate of motion of stars compared to that size, it would seem likely that the Solar System has been comparatively near the center of gravity of the star system for many thousands of years in the past, and will remain comparatively near it for many thousands of years in the future.

Whatever the actual shape of the orbit, a moving Solar System means that the Universe is not likely to be systemocentric by design, but simply by circumstance, and is perhaps not even permanently so.

It is a little uncomfortable to have the Universe of stars seem to possess spherical symmetry and to have that as the only evidence of its systemocentricity. We can't see all the stars, so how do we know they are *really* distributed according to spherical symmetry? It would be nice if there were markings in the sky that could help us in reaching a decision as to systemocentricity or non-systemocentricity.

There *is* such a marking, and a very obvious one. It is the Milky Way, the luminous band of fog that encircles the sky and divides it into two roughly equal halves.

In 1609, the Italian scientist Galileo, turning a small telescope on the sky for the first time, was able to show that the Milky Way was not just a luminous fog, but was a vast crowd of very dim stars, stars that were too numerous and too individually dim to be made out as stars without a telescope.

Why should there be so many stars seen in the direction of the Milky Way, and so few (comparatively) elsewhere?

As early as 1742, an English astronomer, Thomas Wright (1711–86), suggested that the star system was not spherically symmetrical, using the Milky Way as the core of his reasoning.

In 1784, Herschel (who was later to demonstrate that the Sun was moving) decided to check the asymmetry of the Universe by straight-forward observation. It was clearly impractical to try to count *all* the stars. Instead, he chose 683 small patches of equal size that were distributed evenly over the sky and counted all the stars visible to his telescope in each one. In a very real sense, he polled the sky.

He found that the number of stars per patch rose steadily as he approached the Milky Way, was maximal in the plane of the Milky Way, and was minimal in the direction at right angles to that plane.

It seemed to Herschel that the easiest way of explaining this was to suppose that the star system was *not* spherical, but was, instead, shaped like a lens (or a hamburger patty). If we sighted along the long diameter of the lens, we would see more stars than we would in any other direction. We would see so many, in fact, that they would melt together to form the foggy Milky Way. As we looked farther and farther away from the plane of the Milky Way, we would look through a shorter and shorter length of star-filled space and, therefore, see fewer and fewer stars.

Herschel called this lens-shaped star system the "Galaxy" from Greek words for "Milky Way."

If the Solar System were far from the central plane marking out the long diameters of the Galaxy, we would see the Milky Way as a circle of light confined to one side of the sky. It would look like a doughnut, with the stars more thickly strewn in the hole of the doughnut than in the wide spaces outside the doughnut. The farther we were to one side of the plane, the smaller the circle of light marking the doughnut, the more thickly strewn the stars within, and the less thickly strewn the stars outside.

As it happens, though, the Milky Way divides the sky into two halves, with stars equally spread in each half. This is rather conclusive proof that we are in or very near the central plane of the Galaxy.

Even though we might be in the central plane of the Galaxy, we might be far from the actual central point of that plane. If we were, then the Milky Way would be thicker and more luminous in one half of its circle than the other. The farther we were from the central point, the greater the asymmetry in this respect.

As it happens, though, the Milky Way is reasonably equal in width and luminosity all around the sky so that the Solar System must be at, or quite near, the center.

The Galaxy, then, would seem to be systemocentric, and since, in Herschel's time and for a century afterward, the Galaxy was thought by most astronomers to comprise all the stars in the Universe, the Universe itself must be systemocentric.

This view was retained as late as 1920, when the Dutch astronomer Jacobus Cornelis Kapteyn (1851–1922) estimated the Galaxy (and Universe) to be 17,000 parsecs wide and 3,400 parsecs thick, with the Solar System near the center.

Yet this was all wrong. The Solar System was no more at the center of the Galaxy (despite the evidence of the Milky Way) than Earth was at the center of the planetary system.

How that was discovered will be explained in the next chapter.

XIV

Out in the Boondocks

In 1584, the French satirist François Rabelais wrote, "Everything comes to him who knows how to wait." This has been repeated in one form or another ever since, so that Disraeli and Longfellow are among those who are independently quoted to that effect. The aphorism is best known today in the slightly shorter form: "Everything comes to him who waits."

I have never been overly impressed by this comment, however. I felt that for many things one might have to wait far longer than one was likely to live. After all, please note that all the aphorism-makers carefully refrain from putting an upper limit on the period of waiting.

In my own case, it seemed to me (quite early in the game) that I would never have a book on the best-seller lists, no matter how skillfully I might wait.

Mind you, this is not to say that my books don't sell well. Some do. In fact some sell very well—but only over the course of years and decades. They never sell *intensively*. They never sell so much in any one particular week as to earn a place on the New York *Times* best-seller list.

I grew reconciled to that. I even labored to convince myself that this was the result of my integrity and virtue.

After all, my books never concern themselves with sex in clinical detail, or with violence in unpleasant concentration, or, indeed, with any form of sensationalism. On the positive side, they tend to be cerebral, with great emphasis on the rational discussion of motives and of alternate courses of action. Obviously, this, if well done, would have great appeal to a relatively small number of readers.

These few, I was well aware, would be several cuts above the average in intelligence, and would be intensely loyal. These were *my* readers, and I loved them, and I would not change them for a trillion of the more ordinary kind.

And yet sometimes, in the middle of the night, when I was alone in the deepest recesses of my mind, I would wonder what might happen if, just for a little while, *everyone* was above average in intelligence, so that one of my books would—just once—for just one week, be on the best-seller list.

Then I would dismiss the thought as pure fantasy.

In this way it came about that by October 1982, I had been a professional writer for forty-four years, and had published 261 books, without one best seller on the list. I had long decided that there was a kind of distinction about this that I ought to feel proud of. How many other writers, after all, could write 261 books with such unerring failure to hit the target?

And then it came to pass that on October 8, 1982, Doubleday published my 262nd book, which was *Foundation's Edge*, the fourth volume of my Foundation series. This came thirty-two years after I had written what I had decided would be the last word of the series. During all that time, I had continued to turn a resolutely deaf ear to the pleadings of my readers and editors for more. (Well, *they* kept waiting, and it came—as good old François had told them it would.)

As my editor, Hugh O'Neill, had staunchly predicted from the start, the book immediately hit the best-seller lists. On October 17 there appeared the Sunday New York *Times* on my doorstep, and there, on the list in the Book Review section, in bold letters, was *Foundation's Edge* by Isaac Asimov.

After forty-four years, my 262nd book hit the target, even though it was just as nonsexual, nonviolent, nonsensational, and as thoroughly cerebral as all the rest—if not more so. It had been necessary only to wait.

Doubleday hosted a lavish party in my honor and, for a dazzled while, I felt the center of the Universe—which brings me back to the matter under discussion in the previous chapter.

In that previous chapter, I dealt with the natural yearning of people to be the center of the Universe. At first each person thought himself to be the center, and then that post was abandoned (reluctantly) to some site of cultural importance, then to the Earth as a whole, and then to the Solar System as a whole.

Even as late as the 1910s, it seemed reasonable to suppose that the Solar System was at or near the center of the Galaxy (and the Galaxy was then suspected of being just about the entire Universe).

After all, the various objects in the sky seemed to be placed symmetrically about us. Thus, the stars are not more thickly spread in one half of the sky than in the other; and the Milky Way, which represents the Galaxy as viewed through its long diameter, divides the sky into two more or less equal halves.

Before there is good reason to suspect that we are not more or less central in our position, some indisputable asymmetry must be discovered in the sky.

And one exists. The story of that asymmetry begins with Charles Messier, who specialized in comets. He was one of those who early spotted Comet Halley on its return in 1759, the return that had been predicted by Edmund Halley himself (see Chapter 10).

After that, Messier took off. In the next fifteen years he made almost all the comet discoveries that took place, twenty-one of them by his own count. It was the passion of his life, and when he had to attend his wife on her deathbed, and thus missed discovering a comet, which was announced by a competing French astronomer instead, Messier is believably reported to have wept over the lost comet and all but ignored his lost wife.

What particularly bothered Messier was that every once in a while in his search for some tiny fuzzy object in the sky that would indicate the presence of a distant comet heading inward toward the neighborhood of the Sun, he would come across some tiny fuzzy object which, as it happened, was *always* present in the sky. He hated seeing one of the latter, growing excited, then being disappointed.

Between 1774 and 1784, he began to make, and publish, a list of 103 objects which, he felt, should be memorized by serious comet hunters who would, in this way, never be misled into mistaking something worthless for something of cometary importance. The objects on his list are still known as "Messier 1," "Messier 2," and so on (or "M1," "M2," and so on.)

And yet, as it happened, his comet discoveries are trivial, while the objects he listed, in order that astronomers might learn to ignore them, proved of first-rate importance. The very first object on his list, for instance, happens to be the most important single object in the sky beyond the Solar System—the Crab Nebula.

Another object on Messier's list, M13, was first reported in 1714 by none other than Halley, the patron saint of all comet hunters.

In 1781, William Herschel received a copy of Messier's list. It was his ambition to examine every object in the sky, so he made up his mind to look at each item on the list, including, of course, M13.

Herschel, unable to afford a good telescope when he first became interested in astronomy, set about making his own, and ended by making the best telescopes of his time. The telescope he used to look at the Messier objects was far better than those available to Halley or Messier, and when Herschel looked at M13, he saw not just fuzz, as the two earlier astronomers had, but a densely packed spherical conglomeration of stars.

Herschel was the first to interpret correctly the nature of what we now call "globular clusters." Since M13 is in the constellation Hercules, it is sometimes called the "Great Hercules Cluster." Herschel discovered other globular clusters as well, and it turned out that about a quarter of all the objects on Messier's list were globular clusters.

These clusters are made up of hundreds of thousands of stars, the larger ones containing, possibly, millions of them. The star density within these clusters is enormous. At the center of a large cluster of this type there may be as many as 1,000 stars per cubic parsec, while in our own neighborhood there is something like 0.075 stars per cubic parsec.

If we were at the center of a large globular cluster (and could survive there) we would see a night sky filled with a total of about 80,000,000 visible stars, of which (if the luminosity distribution there were what it is here) over 250,000 would be first magnitude or better.

Yet the globular clusters are so far away that the conglomeration of all those stars form units that are only in a few cases visible from Earth to the unaided eye, and then just barely.

What is most interesting about the hundred or so globular clusters that are now known, however, is that almost all of them are on one side of the sky, almost none of them on the other. Nearly one third of them are to be found within that portion of the sky subtended by the single constellation Sagittarius. This asymmetry was first noticed by Herschel's son, John (1792–1871), a noted astronomer in his own right.

This is the most remarkable asymmetry we can observe in the sky, yet it is not in itself sufficient to shake the suggestion that the Solar System is at the center of the Galaxy. There's just a chance, after all, that it might all be a coincidence; that the globular clusters might just happen to be all on one side of us.

A turning point came in 1904, when the American astronomer Henrietta Swan Leavitt (1868–1921) first established a relationship between the length of the period of a type of star called a "Cepheid variable" and its intrinsic brightness, or "luminosity" (see "The Flickering Yardstick," in *Fact and Fancy*, Doubleday, 1962).

This meant it was possible, in principle, to compare the luminosity of a Cepheid variable with its apparent brightness in the sky, and to judge its distance from that—a distance that might be too great to judge in any other way then known.

In 1913, the Danish astronomer Ejnar Hertzsprung (1873–1967) converted this potentiality into reality, and was the first to estimate the actual distances of some Cepheid variables.

This brings us to the American astronomer Harlow Shapley (1885–1972), who had earned his education with great difficulty because of his poverty-stricken childhood, and who became an astronomer by accident. He had entered the University of Missouri in order to become a journalist, but their School of Journalism was not scheduled to open for a year, so young Shapley took a course in astronomy just to fill in the time—and never became a journalist.

Shapley grew interested in Cepheid variables and, by 1913, had demonstrated that they were not binary stars that eclipsed each other. He suggested, instead, that they were pulsating stars. About ten years later, the English astronomer Arthur Stanley Eddington (1882–1944) worked out the theory of Cepheid pulsations in great detail, and settled the matter.

Once Shapley joined Mount Wilson Observatory in 1914, he undertook to investigate variable stars in globular clusters. In doing so, he discovered that they contained stars of a kind called "RR Lyrae variables," because the best-known example of that class was a star known as RR Lyrae.

The manner in which the light of an RR Lyrae variable increases and diminishes is just like that of a Cepheid variable, but the period of variation of the former is smaller. RR Lyrae variables usually have a period of less than one day, whereas Cepheid variables have a period of a week or so.

Shapley decided that the difference in period of variation was not significant and that RR Lyrae variables were simply short-period Cepheid variables. He felt, therefore, that the relationship between brightness and period worked out by Leavitt for the Cepheid variables would work for the RR Lyrae variables as well. (In this, as it turned out, he was right.)

He proceeded to record the brightness and period of RR Lyrae variables in every one of the 93 globular clusters then known, and that gave him, at once, the *relative* distance of these clusters. Since he knew the direction in which they were located, and had determined their relative distances, he could build a three-dimensional model of their distribution.

By 1918, Shapley had demonstrated to his own satisfaction (and, soon, to that of astronomers, generally) that the globular clusters were distributed with spherical symmetry about a point in the plane of the Milky Way, but a point very far from the Solar System.

If the Solar System was at or near the center of the Galaxy, it meant that the globular clusters were centered about or beyond one end of the Galaxy. Their maldistribution in the skies of Earth would then be indicative of their actual asymmetric distribution with respect to the Galaxy.

This didn't seem to be sensible, however. Why should these vast clusters of stars find something so interesting at one end of the Galaxy when all our experience with the way the Law of Universal Gravitation works should lead us to believe that the clusters would be symmetrically distributed about the *center* of the Galaxy?

Shapley made the dramatic decision that the globular clusters *were* distributed about the center of the Galaxy, and that what we thought of as one end of the Galaxy *was*, in point of fact, the center of the Galaxy, and it was *we*, not the globular clusters, who were at one end of it.

But if that was so, it became necessary to explain the symmetry of everything else in the sky. If we were far at one end of the Galaxy, and if the center lay in the direction of Sagittarius, where the globular clusters were most concentrated, then why did we not see far greater numbers of stars in the Sagittarius direction than in the opposite Gemini direction? Why wasn't the Milky Way in Sagittarius far brighter than in Gemini?

Such questions had to be answered; all the more so as confirmatory evidence of Shapley's suggestion quickly arose.

In the 1920s, the "spiral nebulae," observed here and there in the sky, were found to be not masses of gas, as had been suspected, but vast assemblages of stars; they were galaxies in their own right.

The nearest spiral galaxy is in the constellation of Andromeda, and a study of this Andromeda Galaxy showed that it, too, possessed globular clusters, just like our own if we allowed for the much greater distance of those of the Andromeda Galaxy.

The globular clusters of the Andromeda Galaxy were distributed with spherical symmetry about that galaxy's center, just as Shapley said our own Galaxy's globular clusters ought to be. We could *see* that this was the way in which the Andromeda Galaxy's globular clusters behaved, and there was no reason to believe that ours should behave any differently.

It came to be accepted, therefore, and was eventually demonstrated beyond a reasonable doubt, that our Milky Way Galaxy is a spiral galaxy much like the Andromeda Galaxy, and that the Solar System is not in its center, but is far out in one of the spiral arms.

Humanity, the Earth, the Sun, the entire Solar System is not near the center of things with respect to our Galaxy. Not at all! We are in the galactic suburbs, out in the boondocks. That may be humiliating, but that's the way it is.

To be sure, we *are* in or near the galactic plane. That is why the Milky Way cuts the sky into two equal halves.

But the symmetry! Why is the Milky Way more or less equally bright all around?

If we examine the Andromeda Galaxy, and other spiral galaxies close enough to be seen in some detail, we find that the spiral arms are rich in dust clouds that do not enclose stars and that are therefore not illuminated. They are "dark nebulae."

If such dark nebulae existed in space far from any stars, they would not be seen. They would be black-on-black, so to speak. If, on the other hand, there were clouds of stars behind the nebulae, the dust particles in the nebulae would prove efficient absorbers and scatterers of the light behind them, and observers would see the clouds as dark masses against the starlight present on every side.

The spiral arms of our own Galaxy are no exception to this.

William Herschel, himself, in his indefatigable study of everything in the sky, noticed places in the Milky Way where there were inter- ruptions, quite sharply marked off, in the smooth distribution of the numerous faint stars, regions where there were no visible stars at all. Herschel thought that these were regions that lacked stars in truth, and that these tubes of nothingness, reaching through what, to Herschel, seemed a rather thin depth of stars in the Milky Way, were so oriented that we could look through them. "Surely," he said, "this is a hole in the heavens."

More and more of these regions were found (the number now comes to over 350), and it seemed increasingly unlikely that there were so many starless holes in the heavens. About 1900, the American astron- omer Edward Emerson Barnard (1857–1923) and the German astron- omer Max Franz Joseph Cornelius Wolf (1863–1932) independently suggested that these Milky Way interruptions were dark clouds of dust and gas that obscured the light of the numerous stars behind them.

It was these dark nebulae that explained the symmetry of the Milky Way. So clogged was the Milky Way with them that the light from the central regions of the Galaxy, and from the spiral arms beyond the center, is totally obscured. All we can see from Earth is our own neighborhood of the spiral arms of the Galaxy. We can see about equally far into the Milky Way in all directions, so that what we *see* of the sky is symmetrical.

Shapley not only worked out the relative distance of the globular clusters, but he also devised a statistical system for treating the RR Lyrae variables in such a way as to estimate the absolute distance from the Earth of the globular clusters. Shapley's system was legitimate, but there was a factor he didn't take into account, and that led him to overestimate the size of the Galaxy.

Again, it was a matter of light being obscured, even when dark nebulae were absent.

There is an analogy to this in the case of Earth's atmosphere. Atmospheric clouds in the sky can obviously obscure the Sun, but even the "clear" air of a cloudless sky is not *completely* transparent. Some light is scattered and absorbed. This is particularly noticeable near the horizon where light must travel a far greater thickness of atmosphere to reach our eyes or our instruments. Thus the Sun at the horizon has its rays so enfeebled that we can often look at it with impunity, and, as for stars, they can be dimmed to invisibility.

Similarly, there are sparsely spread-out atoms, molecules, and even dust particles in "clear" space. Space is, of course, far clearer than our atmosphere, even at the latter's clearest, but starlight must travel across many trillions of kilometers to reach us, and over such a distance even very occasional bits of matter will produce cumulative effects that are noticeable.

This was made clear, in 1930, by the Swiss-American astronomer Robert Julius Trumpler (1886–1956), who demonstrated that the brightness of star clusters fell off with distance a little more rapidly than would be expected if space were completely clear. He therefore postulated the existence of exceedingly thin interstellar matter, and this has been amply demonstrated since.

The presence of such dust in "clear" space, something Shapley did not allow for, dims the RR Lyrae variables in the globular clusters, so that one calculates them a bit farther away than they really are. Once the Trumpler correction was introduced, the size of the Galaxy was reduced somewhat from Shapley's estimate, and the values thus found are still accepted today.

At present, the Galaxy is considered to be a vast lens-shaped (or hamburger patty-shaped) object, which, if seen in cross section, is very wide across and relatively narrow up and down.

The long diameter is about 30,000 parsecs (or about 100,000 light-years, or about 30 quadrillion kilometers). It is about 5,000 parsecs thick at the center, and about 950 parsecs thick out here where the Solar System is. —For comparison, the nearest star, Alpha Centauri, is about 1.3 parsecs away from us, and if it (or our Sun) were 15 parsecs away it would be barely visible to the unaided eye.

From the center of the Galaxy to its outer perimeter the distance is about 15,000 parsecs, and we are some 9,000 parsecs from the center. We are thus more than halfway from the center to the outer perimeter, which is about 6,000 parsecs from us in the direction away from the center.

From our study of other galaxies, we have discovered, in the last quarter century or so, that galactic centers are unexpectedly violent places. They are so violent, in fact, that it seems likely that life as we know it is completely impossible in the central regions of galaxies, and is likely to exist only in the boondocks, where we are.

It is important to study all that violence from a safe distance, for a better understanding of what goes on might tell us a great deal about the Universe that we could not work out otherwise. Astronomers are doing their best to do so. The trouble is that the distances to the centers of other galaxies is entirely too great. We could afford to be closer and still be quite safe.

The center of the nearest giant galaxy, the Andromeda Galaxy, is about 700,000 parsecs away, for instance. The only comparable region that is any closer is the center of our own Milky Way Galaxy, which is only 9,000 parsecs away, less than $1/80$ the distance of the Andromeda Galaxy's center. The only trouble is that we can't see the center of our own Galaxy, close as it is.

But, wait, when I say we can't see it, I mean by visible light, because it is permanently fogged in by galactic dust.

On Earth, however, when clouds or fog obscures the view, we can use radar. The short-wave radio beams emitted and received by our radar devices can penetrate clouds and fog without trouble.

As it happens, astronomical objects that are capable of emitting light are also capable of emitting radio waves, and at times these radio waves are emitted with great intensity. Such radio waves, unlike light waves, can penetrate great clouds of dust without trouble.

In 1931, Karl Jansky first detected radio waves in the sky. Those radio waves might have come from the Sun, which, when it is at or near the peak of its sunspot activity, is the strongest radio source in the sky (because it is incredibly near, as stellar distances go). The Sun, however, happened to be in a quiet stage, so that Jansky picked up the next strongest source, which was a spot in Sagittarius.

Of course, Sagittarius is the direction of the galactic center, and there is no question at all but that the intensely energetic beam of radio waves which Jansky detected is coming from that center.

With present-day radio telescopes, one can zero in exactly on the location of the source, and it has now been narrowed down to a spot no wider than 0.001 seconds of arc.

This is amazingly tiny. The planet Jupiter, when nearest to us, is 3,000 seconds of arc across, so that the central galactic radio source is only 1/3,000,000 as wide as Jupiter appears to be in our sky—and Jupiter appears only as a dot of light.

Of course, the central source is enormously farther away than Jupiter is, and if we make allowance for that distance, the width of the central source would appear to be about 3,000,000,000 kilometers. If the central source were transferred (in imagination) to the position of our Sun, it would be seen to be the size of an enormous red giant star, filling all of space out to the orbit of distant Saturn.

Yet large as that is on the scale of the Solar System, it is far from large enough to account for the energy that pours out of it. An ordinary star like our Sun radiates at the expense of nuclear fusion, but no reasonable amount of nuclear fusion can be packed into something the size of the central source and produce the amount of energy it appears to produce.

The one energy source that is still more efficient is gravitational collapse. The growing opinion, therefore, is that at the center of our Galaxy (and, possibly, at the center of all galaxies, and even of all sizable globular clusters) is a black hole.

Our own galactic black hole may have a mass of a million times that of our Sun. It should be steadily growing, swallowing matter out of the rich concentration existing at the core of the Galaxy (where the stars are even more densely distributed than at the core of a globular cluster) and converting part of that mass into the energy it radiates.

Larger galaxies would have more massive black holes, radiating even more energetically as they gobbled matter. Active galaxies, such as the Seyfert galaxies (first noted in 1943 by the American astronomer Carl Keenan Seyfert [1911–60]), must have still more energetic events taking place in their extraordinarily bright centers. As for quasars, which are increasingly being thought of as super-Seyfert galaxies, the events at their center must be the most violent of all in our present-day Universe.

We could perhaps get an insight into all these violences and super-violences if we study the not-so-distant center of our own Galaxy in detail—a center the very existence of which we did not even suspect until some sixty years ago.

MATHEMATICS

XV

To Ungild Refinèd Gold

One of my less amiable characteristics is an impatience with misquotation, especially if it is from Shakespeare.

A sure way to induce the symptoms of apoplexy in me is to have someone who is doing a comic rendition of the passage from *Romeo and Juliet* say "Wherefore art thou, Romeo?" with an intonation and action indicating that the meaning is "Where are you, Romeo?"

Not only does this indicate that the poor illiterates who are responsible have never read the play but that they don't even know the meaning of "wherefore," or—worse yet—assume that the audience neither knows nor cares.

A high follower in the list of misquoters whom I disesteem are those who speak of "gilding the lily."

That is a misquotation from Shakespeare's *King John*, Act IV, Scene ii, where the Earl of Salisbury lists six actions that represent "wasteful and ridiculous excess" as a way of condemning King John's insistence on a second coronation. In each case, something is described that attempts to improve that which cannot be improved, and the first two examples are "to gild refinèd gold, to paint the lily."

The misquoter collapses the two and says "to gild the lily," an action which somewhat lacks the exquisite inappropriateness of the two actions as given by Shakespeare.

So as my way of fighting this annoyance I intend to demonstrate a way in which I can manage "to ungild refinèd gold." You'll see what I mean as I go on.

When I am trapped in an assemblage, and am restless, and am sure no one is eying me closely, I can sometimes rescue myself by playing with numbers: adding, subtracting, multiplying, dividing, and so on.

There is no point to my doing this, you understand, because I lack all trace of mathematical talent. What I do with numbers is to mathematics what piling one toy block on top of another is to architecture. But then, you see, I don't imagine myself to be doing mathematics; I am merely protecting my brain (a rather demanding organ) from damage through boredom.

I did this even when I was quite young, and I was about twelve, I think, when I studied the relation of numbers to their squares in the following fashion:

$$1^2 = 1; \text{ and } 1 - 1 = 0$$
$$2^2 = 4; \text{ and } 4 - 2 = 2$$
$$3^2 = 9; \text{ and } 9 - 3 = 6$$
$$4^2 = 16; \text{ and } 16 - 4 = 12$$
$$5^2 = 25; \text{ and } 25 - 5 = 20$$
$$6^2 = 36; \text{ and } 36 - 6 = 30$$

By this time, I saw the regularity. If you go up the scale of integers, subtracting each integer from its square, the first integer will give you 0. You must then add 2 to get 2 for the next integer; add 4 to get 6 for the next integer; add 6 to get 12; add 8 to get 20; add 10 to get 30.

In producing the successive numbers, you go up the scale of even integers, so that I knew that the next number would be 42, and then 56, and then 72, without having to carry out the subtractions: $49 - 7$; $64 - 8$; and $81 - 9$. —I was very proud of myself.

I next tried something else. I wrote down each integer, and placed next to it the figure I got by subtracting it from its square and then considered how else I could represent the figure. Thus:

$$1 \underline{} 0 = 1 \times 0$$
$$2 \underline{} 2 = 2 \times 1$$
$$3 \underline{} 6 = 3 \times 2$$
$$4 \underline{} 12 = 4 \times 3$$
$$5 \underline{} 20 = 5 \times 4$$
$$6 \underline{} 30 = 6 \times 5$$

It was clear to me that every integer subtracted from its own square gave a result that was equal to that same integer multiplied by the next smaller integer.

By now, my twelve-year-old heart was beating quickly, for I got the idea that I had discovered something very unusual that, perhaps, no one had ever noticed before. (As I told you, I have no mathematical talent. A real mathematician would have noticed all this at the age of three, I imagine, and dismissed it as obvious.)

At any rate, I wanted to generalize this for I was taking algebra now. I therefore let an integer, *any* integer, be "x." The next smaller integer would be "x − 1," and the square of the integer would be "x^2."

It turned out, then, I had discovered, by massive brain power, that an integer subtracted from its square, "$x^2 - x$," was equal to that integer multiplied by the next smaller "x(x − 1)." In other words:

$$x^2 - x = x(x - 1)$$

With that, all my joy left me, for this equation was indeed as obvious as anything could be. I just factored out the "x" on the left side, and that gave me the right side. The value of my discovery was equal to that of finding out that two dozen equaled twenty-four.

I therefore abandoned that particular line of discovery and never returned to it. —And this was a shame, for had I continued to pry, I might conceivably have discovered something which, while not exactly new, would have been far more interesting than the equation I just worked out for you. And since I'm a little over twelve now, I can manage it—so come along.

Suppose we consider the problem of subtracting an integer from its square for just the first three integers: 1 − 1 = 0; 4 − 2 = 2; and 9 − 3 = 6. The differences keep going steadily upward so that we can tell that there are no subtractions of this sort that will ever give differences of 1, 3, 4, or 5. At least, not if we stick to integers.

We might, however, switch to the use of decimal fractions.

For instance, the square of 1.1 is 1.21, and $1.21 - 1.1 = 0.11$, while the square of 1.2 is 1.44, and $1.44 - 1.2 = 0.24$. If we continue going upward by tenths, we find that the square of 1.6 is 2.56, and $2.56 - 1.6 = 0.96$, which is quite close to 1. Then, too, the square of 2.3 is 5.29, and $5.29 - 2.3 = 2.99$, which is even closer to 3.

In fact, we can guess, by now, that if we choose the proper decimal fraction, we can subtract that from its square and get a number pretty close to any integer we choose. Thus, the square of 4.65 is 21.6225, and $21.6225 - 4.65 = 16.9725$, which is pretty close to 17.

None of the examples I have cited give a difference that is an exact integer; they only come close. My twelve-year-old self, if he were working this out, and were as clever as I wish he had been, might have thought that by adding more decimal places he could have hit an integer on the nose. Since $2.3^2 - 2.3 = 2.99$, it should seem reasonable to expect that a tiny upward adjustment of 2.3 would get me exactly 3. For instance, $2.303^2 - 2.303 = 3.000809$. Now I'm just a trifle high, so down I go to $2.30275^2 - 2.30275 = 2.9999075$.

When I was twelve, I would not have had a pocket calculator, so working out the above relationships would have taken me quite a time, riddled me with arithmetical errors, and worn me out. I would have quickly given up.

Suppose, however, I didn't. Suppose I had had the guts and persistence to try more and more decimal places and to fill more and more pieces of paper with enormous calculations. I would have found that no matter how assiduously I tried, and how many hours (or years) I spent at it, I would never find any number with any quantity of decimal places which, when subtracted from its square, would give exactly 3. I would get closer and closer and closer, but nothing would place me exactly on 3.

There would then be two possible conclusions I could have come to: (1) If I had been an ordinary boy, I would have decided that I just lacked the persistence to reach that final decimal place. (2) If I had been a boy with the soul of a mathematician, I would have leaped intuitively to the notion that the number I was after was, actually, an unending and nonrepeating decimal, and I might thus have caught my first glimpse—unaided—of irrational numbers. (Unfortunately, I was never even bright enough to get to the point where I had to make the choice, so I was subordinary.)

As I went on in algebra, I discovered how to solve for "x" in equations of the following type: "$ax^2 + bx + c = 0$." In such an equation, "a," "b," and "c," the "coefficients," are integers, and "x" is unknown. It turns out that in such an equation:

$$x = (\tfrac{1}{2}a)(-b + \sqrt{b^2 - 4ac}) \text{ —— Equation 1}$$

A couple of explanations: In Equation 1, the quantity in one parenthesis is to be multiplied by the quantity in the other parenthesis. Then, too, the symbol $\sqrt{}$ stands for "square root." The "square root of x," or "$\sqrt{x}$," is that number which, when multiplied by itself, gives "x." Thus, $\sqrt{25} = 5$, $\sqrt{81} = 9$, and so on. (One more point: If the plus sign in Equation 1 is replaced by a minus sign, a second possible answer would be given, but we will deal with the plus sign only.)

To give an example of how Equation 1 works, suppose we deal with an equation such as "$x^2 + 8x - 5$." In that case "c" is equal to -5, and "b" is equal to $+8$. However, the plus sign is usually omitted in such cases, and is considered to be "understood," so that "b" is said to be equal simply to 8.

But what is "a," the coefficient of "x^2," in the equation: "$x^2 + 8x - 5$"? It would seem that the "x^2" in that equation has no coefficient at all, but that is not so. The "x^2," standing by itself, is actually "$1x^2$," but the 1 is understood and is generally omitted. Nevertheless, "a," in this case, is set equal to 1. (Personally, I would never omit anything and would always write 8 as $+8$, and "x^2" as "$1x^2$," and for that matter "x" as "x^1," but that's not what mathematicians do. It's their amiable way of saving themselves trouble at the expense of making things a little more confusing for beginners, and you can't fight Faculty Hall.)

We are now ready to return to the matter of subtracting an integer from its square to get some desired difference. We can generalize the problem algebraically by letting "x" stand for any integer, "x^2" for its square, and "y" for the integer that is the difference. We would then write:

$$x^2 - x = y$$

To make it interesting, let's pick a specific integer for "y," so we can see how this works; and, to make it simple, let's pick the smallest integer, 1. The equation becomes:

$$x^2 - x = 1$$

It is possible to subtract 1 from each side of the equality sign without changing the nature of the equation. (That's the result of one of those good old axioms: Equals subtracted from equals are equal.)

If you subtract 1 from the left side of the equality you get "$x^2 - x - 1$." If you subtract 1 from the right side, you get $1 - 1$, which is equal to 0. So you can write the equation as:

$$x^2 - x - 1 = 0 \text{ —— Equation 2}$$

If you solve this equation for "x," you will have a number which, when subtracted from its square, will give you exactly 1.

For the purpose, we will use Equation 1. For "a," the coefficient of "x^2," we have 1, for "b," the coefficient of "x," we have -1, and for "c," the final coefficient, we have -1 again.

Since "b" is equal to -1, "$-b$" = $-(-1)$, or $+1$, which is written simply 1. Again "b^2" = $(-1)(-1)$, or $+1$, or 1. Since "a" is equal to 1, then "$\frac{1}{2}a$" is equal to $\frac{1}{2}$. And since "a" = 1, and "c" = -1, "4ac" is equal to $4(1)(-1)$ or -4, and $-4ac$ is equal to $-(-4)$, or $+4$, or 4.

With all this in mind, we have all we need to know to substitute numbers for the symbols in Equation 1 (and forgive me if you didn't require this step-by-step explanation). Equation 1 therefore becomes:

$$x = \frac{1}{2}(1 + \sqrt{1 + 4}) = \frac{1}{2}(1 + \sqrt{5})$$

This is the number which, when subtracted from its square, will yield a difference of exactly 1.

To express the number as an ordinary decimal, you must take the square root of 5, add 1, and divide the sum by 2.

But what is the square root of 5? What is the number which, when multiplied by itself, will give 5? That, alas, is an irrational number, an unending and nonrepeating decimal. We can get pretty close, though, if we say it is 2.23606796. . . . In fact, we'll be close enough if we pretend that the square root of 5 is 2.236068. If we multiply this number by itself, 2.236068×2.236068, we get 5.0000001, which is off by only a ten-millionth.

If we add 1 to the square root of 5 and divide the sum by 2, we get 1.618034. (A still more correct value would be 1.61803398 . . . but 1.618034 is quite good enough for our purposes.)

If you take the square of this number, you find that $1.618034 \times 1.618034 = 2.618034$, and the difference is 1.

In fact, it's not quite so accurate: $1.618034 \times 1.618034 = 2.618034025156$. The additional 0.000000025156 is the result of the trifling inaccuracy of the figure 1.618034. No decimal, however long, could be anything but a trifle inaccurate. The only truly accurate figure is $\frac{1}{2}(1 + \sqrt{5})$. If you square *that* number, which can be done without much trouble, but which trouble I will spare you, you will get the quantity $\frac{1}{2}(3 + \sqrt{5})$, which is greater by *exactly* 1.

Suppose we consider "reciprocals" next. If you divide 1 by any number, you get another number that is the reciprocal of the first. In other words, $\frac{1}{2}$ is the reciprocal of 2; $\frac{1}{3}$ is the reciprocal of 3; 1/17.25 is the reciprocal of 17.25; and, in general, "1/x" is the reciprocal of "x."

Instead of subtracting a number from its square, let's subtract a reciprocal from its number. Using only integers, we have:

$$1 - \tfrac{1}{1} = 0$$
$$2 - \tfrac{1}{2} = 1\tfrac{1}{2}$$
$$3 - \tfrac{1}{3} = 2\tfrac{2}{3}$$
$$4 - \tfrac{1}{4} = 3\tfrac{3}{4}, \text{ and so on.}$$

Except for the case of 1, we would always get a fraction, but, again, we do not have to cling to integers. Suppose we want to find a number which, when we subtract its reciprocal, will give us a difference of exactly 1.

Clearly, it will have to be a number which lies between the integers 1 and 2, so that the difference will be somewhere between 0 and $1\frac{1}{2}$. Suppose, for instance, we take the number 1.5. Its reciprocal is 1/1.5. Since $1.5 = \frac{3}{2}$, and $1/1.5 = \frac{2}{3}$, we have $\frac{3}{2} - \frac{2}{3} = \frac{5}{6}$, which is pretty close to 1. If we move to 1.6 and subtract 1/1.6, and if you trust me with the arithmetic, the answer is 0.975, which is even closer.

If we keep experimenting, however, we will quickly assure ourselves that we're not going to find any decimal we can write down which will give us *exactly* 1, when its reciprocal is subtracted. We are going to find ourselves in the realm of irrational numbers again.

So we switch to algebra and set up an equation that will represent the general case:

$$x - 1/x = 1$$

If we multiply each side of the equation by "x," the nature of the equation is left unchanged (trust me!). Since "x" times "x" is "x^2"; "1/x" times "x" is 1; and 1 times "x" is "x," we have:

$$x^2 - 1 = x$$

If we subtract "x" from each side of the equation, we have:

$$x^2 - 1 - x = 0, \text{ or, rearranging}$$
$$x^2 - x - 1 = 0$$

But this is Equation 2, again, and the solution for "x" must be the same as it was before. We already know that $\frac{1}{2}(1 + \sqrt{5})$ is exactly 1 less than its square. Well, it is also exactly 1 more than its reciprocal.

To show this plainly, let's deal with 1.618034, that very good approximation of $\frac{1}{2}(1 + \sqrt{5})$. It turns out that $1/1.618034 = 0.618034$.

Now let's try again. Imagine a rectangle that is 1 unit wide and 2 units long. (It doesn't matter what the units are: inches, meters, light-years, whatever.)

In such a rectangle, the length is 2 times the width, obviously. The length and the width together, however, is 3 units, and that is 1½ times the length.

If the rectangle is 1 unit by 3, then the length is 3 times the width, but the length and width together is 4 and that sum would be 1⅓ times the length.

If the rectangle were 1 by 4, then 1 by 5, and so on, you would get pairs of figures that were 4 and 1¼, 5 and 1¹/₅, and so on. The two numbers would move farther and farther apart in size.

Can we find a rectangle where the two numbers are equal in size?

If so, that would have to be one in which the width was 1 unit and the length was less than 2 units (because at 2 units, the two numbers are already unequal).

Let's go straight to algebra. Suppose the width of the rectangle is 1 unit and the length is "x" units. To express how many times "x" is greater than 1, we divide "x" by 1, and write "x/1," or simply "x."

The sum of the width and the length of the rectangle is "x + 1." To express how many times this is greater than the length alone, we have "(x + 1)/x."

We are looking for a situation in which these two relative lengths, or "ratios," are equal, so the equation we set up is:

$$x = (x + 1)/x$$

If we multiply both sides of the equation by "x," we don't change the nature of the equation and we have:

$$x^2 = x + 1$$

If we subtract "x + 1" from both sides, we don't change the nature of the equation and we have:

$$x^2 - (x + 1) = 0$$

We can remove the parenthesis if we take the negative of "x + 1" and make it "−x − 1" so that we have:

$$x^2 - x - 1 = 0$$

and there's Equation 2 again, with its usual solution.

Suppose, then, we have a rectangle in which the width is 1 unit and the length is 1.618034 units. The width and the length taken together would therefore be 2.618034. The length is, of course, 1.618034 times the width, while the length and width added together would be 2.618034/1.618034, or 1.618034 times the length alone.

It was the ancient Greeks who discovered this. Essentially, it was a way of dividing a given line into two sections, the longer of which was to the shorter section as the whole line was to the longer section. Mathematicians have been so ravished by the beauty of this balance of ratios that, about the mid-nineteenth century, it began to be called the "golden section."

A rectangle in which the width and length represented a line divided by the golden section and bent into a right angle at the division point was called a "golden rectangle."

Many people feel that the golden rectangle represents an ideal configuration that is particularly satisfying from an aesthetic viewpoint. A longer rectangle, they feel, looks too long, and a shorter one too stubby. Therefore, people have sought (and found) examples of golden rectangles in paintings, in statues, in buildings, and in many common artifacts of our society. In books on popular mathematics, the reader is presented with illustrations showing this.

Frankly, I'm skeptical. My own feeling is that aesthetics is a very complicated study and is enormously influenced by the social environment. To try to make much of the golden section in this respect is simplistic. I have only to see films made in the 1920s and 1930s, for instance, to be struck by the extent to which our ideas of female beauty (which one might casually think of as timeless) have changed in one lifetime.

I don't deny the golden rectangle is golden because of the mathematical elegance of the relationship of the sides, but trying to convert this into matters of aesthetics is to gild refinèd gold, and I would like to contribute my own poor bit to ungilding it.

If we cling strictly to mathematics, we find that the golden section is to be found in such simple geometric configurations as the regular decagon (a symmetrical ten-sided figure) and the pentagram (the star you find in the American flag). Particularly interesting in this connection, however, is the Fibonacci series, something I dealt with in an earlier essay in this series (see "T-Formation," in *Adding a Dimension,* Doubleday, 1964). There I merely dealt with some of the large numbers that resulted. Here I will take up another aspect.

The Fibonacci series starts with two 1's and then generates new numbers by making each new one the sum of the two preceding ones.

Thus, if we start the series with 1, 1 . . . , the third number is $1 + 1$, or 2, and that gives us 1, 1, 2. . . . The next number is $1 + 2$, or 3, and now we have 1, 1, 2, 3. . . . There follows $2 + 3$, or 5, so we have 1, 1, 2, 3, 5. . . . Then comes $3 + 5 = 8$, and $5 + 8 = 13$, and so on. The first 21 terms of the Fibonacci series are, therefore:

1, 1, 2, 3, 5, 8, 13, 21, 34, 55, 89, 144, 233, 377, 610, 987, 1597, 2584, 4181, 6765, 10946. . . .

We can continue to add additional terms indefinitely, if we are willing to add larger and larger pairs of terms, but these 21 will be enough for our purposes.

When the Fibonacci series was worked out (by an Italian mathematician named Leonardo Fibonacci [1170–1240]) it involved biological growth. The initial problem dealt with multiplying rabbits, in fact. And yet—

Suppose we consider the ratio of successive numbers in the Fibonacci series, dividing each number by the one before, starting with the second in the series, thus:

$$^1/_1 = 1$$
$$^2/_1 = 2$$
$$^3/_2 = 1.5$$
$$^5/_3 = 1.6666 \ldots$$
$$^8/_5 = 1.6$$
$$^{13}/_8 = 1.625$$

As you see, the ratio forms an oscillating series. The value of the ratio goes up from 1 to 2, then down to 1.5, then up to 1.6666 . . . , then down to 1.6, then up to 1.625. We can be sure that this will go on, that the ratio will continue to move up and down alternately, and, in fact, it does.

However, the ratio goes up and down in successively smaller swings. First it goes from 1 to 2, but in later swings it never gets as low as 1 again, or as high as 2, either. Then it goes down from 2 to 1.5, and all future values are between 1.5 and 2. Then it goes up to 1.6666 . . . and all future values are between 1.5 and 1.6666. . . .

The oscillation gets smaller and smaller and smaller, and with each step all future values are trapped between the smaller and smaller swings.

The oscillation never stops completely. No matter how far we follow out the series, and how immense the numbers get, the ratio will continue to swing, though by ever tinier amounts. This smaller and smaller swing will always be to one side and then the other of some central value, which the ratio will get ever closer to without ever quite reaching. That central value is called the "limit" of the series.

What is the limit of the Fibonacci series?

Let us continue the series, starting with the last ratio we have already dealt with:

$$13/8 = 1.625$$
$$21/13 = 1.6153846 \ldots$$
$$34/21 = 1.6190476 \ldots$$
$$55/34 = 1.617647$$
$$89/55 = 1.6181818 \ldots$$
$$144/89 = 1.6179775 \ldots$$
$$233/144 = 1.6180555 \ldots$$
$$377/233 = 1.6180257 \ldots$$
$$610/377 = 1.6180371 \ldots$$

This is getting to look awfully suspicious. Let's switch to the two last ratios in the Fibonacci series of 21 members that I presented earlier:

$$6765/4181 = 1.618033963 \ldots$$
$$10946/6765 = 1.618033998 \ldots$$

The oscillations are getting very small indeed, and they *seem* to be oscillating about the number that represents the golden section.

Can we be sure? Perhaps they are oscillating about a value that is microscopically different from the golden section.

No, that's not so. There are mathematical methods for determining the limit of such series and one can demonstrate quite conclusively that the limit of the ratios of successive terms of a Fibonacci series is $\frac{1}{2}(1 + \sqrt{5})$.

The fact delights me. It is an example of the beauty of the unexpected that you can find everywhere in mathematics, if you have the talent for it. —As I, alas, have not.

THE FRINGE

XVI

The Circle of the Earth

Once, Janet and I were in a hotel room during the course of one of my lecture engagements, and a chambermaid knocked on the door to ask if we needed any towels. It seemed to me we had towels, so I said that no, we didn't need towels.

I had scarcely closed the door, when Janet called from the bathroom saying that we did, too, need towels, and to call her back.

So I opened the door, called her back, and said, "Miss, the woman I have here in my hotel room with me says we do, too, need more towels. Would you bring some?"

"Sure," said she, and went off.

Out came Janet, with that expression of exasperation she wears whenever my sense of humor escapes her a bit. She said, "Now why did you say that?"

"It was a literally true statement."

"You know you said that deliberately, in order to imply we're not married. When she comes back, you just tell her we're married, you hear?"

Back came the chambermaid with the towels and I said, "Miss, the woman I have here in my hotel room with me wants me to say to you that we're married."

And over Janet's cry of "Oh, *Isaac*," the chambermaid said, haughtily, "I couldn't care less!"

So much for modern morality.

I thought of this incident recently in the aftermath of an essay I wrote for *Science Digest* in which I made the casual statement that the Bible assumes the Earth to be flat.

You'd be surprised at the indignant letters I got from people who denied vigorously that the Bible assumed the Earth to be flat.

Why? After all, the Bible was written in the day when *everyone* assumed the Earth was flat. To be sure, by the time the latest biblical books were written, a few Greek philosophers thought otherwise, but who listened to *them?* I thought it was only reasonable that the men who wrote the various books of the Bible should know no more about astronomy than anyone else at the time and that we should all be charitable and kind to them, therefore.

However, the Fundamentalists are not like the chambermaid in that hotel. When it comes to any suggestion of a biblical flat earth, they couldn't care *more*.

Their thesis, you see, is that the Bible is literally true, every word, and what's more, that it is "inerrant"; that is, that it cannot be wrong. (This follows clearly from their belief that the Bible is the inspired word of God; that God knows everything; and that, like George Washington, God cannot tell a lie.)

In support of this thesis, the Fundamentalists deny that evolution has taken place; they deny that the Earth and the Universe as a whole are more than a few thousand years old, and so on.

There is ample scientific evidence that the Fundamentalists are wrong in these matters, and that their notions of cosmogony have about as much basis in fact as the Tooth Fairy has, but the Fundamentalists won't accept that. By denying some scientific findings and distorting others, they insist that their silly beliefs have some value, and they call their imaginary constructions "scientific" creationism.

At one point, however, they draw the line. Even the most Fundamental of Fundamentalists would find it a little troublesome to insist that the Earth is flat. After all, Columbus didn't fall off the end of the world, and the astronauts have actually seen the world to be a sphere.

If, then, the Fundamentalists were to admit that the Bible assumes a flat Earth, their entire structure of the inerrancy of the Bible falls to the ground. And if the Bible is wrong in so basic a matter, it can be wrong anywhere else, and they might as well give up.

Consequently, the merest mention of the biblical flat Earth sends them all into convulsions.

My favorite letter, arriving in this connection, made the following three points:

1. The Bible specifically says the Earth is round (and a biblical verse is cited), yet despite this biblical statement, human beings persisted in believing the Earth to be flat for two thousand years thereafter.

2. If there seem to have been Christians who insisted the Earth was flat, it was only the Catholic Church that did so, not Bible-reading Christians.

3. It was a pity that only nonbigots read the Bible. (This, it seemed to me, was a gentle remark intended to imply that I was a bigot who didn't read the Bible and therefore spoke out of ignorance.)

As it happened, my letter-writing friend was well and truly wrong on all three points.

The biblical verse he cited was Isaiah 40:22.

I doubt that my correspondent realized it, or would believe it if he were told, but the fortieth chapter of Isaiah begins that section of the book which is called "the Second Isaiah" because it was not written by the same hand that wrote the first thirty-nine chapters.

The first thirty-nine chapters were clearly written about 700 B.C., in the time of Hezekiah, king of Judah, at the time when the Assyrian monarch Sennacherib was threatening the land. Beginning with Chapter Forty, however, we are dealing with the situation as it was about 540 B.C. in the time of the fall of the Chaldean Empire to Cyrus of Persia.

This means that the Second Isaiah, whoever he might have been, grew up in Babylonia, in the time of the Babylonian captivity and was undoubtedly well educated in Babylonian culture and science.

The Second Isaiah, therefore, thinks of the Universe in terms of Babylonian science, and to the Babylonians the Earth was flat.

Well, then, how does Isaiah 40:22 read? In the Authorized Version (better known as the King James Bible), which is *the* Bible to the Fundamentalists, so that every last mistranslation it contains is sacred to them, the verse, which is part of the Second Isaiah's attempt to describe God, reads:

"It is he that sitteth upon the circle of the earth . . ."

There you have it—"the circle of the earth." Is that not a clear indication that the Earth is "round"? Why, oh why, did all those bigots who don't read the Bible persist in thinking of the Earth as flat, when the word of God, as enshrined in the Bible, spoke of the Earth as a "circle"?

The catch, of course, is that we're supposed to read the King James Bible as though it were written in English. If the Fundamentalists want to insist that every word of the Bible is true, then it is only fair to accept the English meanings of those words and not invent new meanings to twist the biblical statements into something else.

In English, a "circle" is a two-dimensional figure; a "sphere" is a three-dimensional figure. The Earth is very nearly a sphere; it is certainly *not* a circle.

A coin is an example of a circle (if you imagine the coin to have negligible thickness). In other words, what the Second Isaiah is referring to when he speaks of "the circle of the earth" is a flat Earth with a circular boundary; a disk; a coin-shaped object.

The very verse my correspondent advanced as proof that the Bible considered the Earth to be a sphere is the precise verse that is the strongest evidence that the Bible assumes the Earth to be flat.

If you want another verse to the same effect, consider a passage in the Book of Proverbs, which is part of a paean of praise to personified Wisdom as an attribute of God:

"When he prepared the heavens, I was there: when he set a compass upon the face of the depth." (Proverbs 8:27).

A compass, as we all know, draws a circle, so we can imagine God marking out the flat, circular disk of the world in this fashion. William Blake, the English artist and poet, produced a famous painting showing God marking out the limits of the Earth with a compass. Nor is "compass" the best translation of the Hebrew. The Revised Standard Version of the Bible has the verse read, "When he established the heavens, I was there, when he drew a circle on the face of the deep." That makes it clearer and more specific.

Therefore, if we want to draw a schematic map of the world as it seemed to the Babylonians and Jews of the sixth century B.C. (the time of the Second Isaiah) you will find it in Figure 1. Although the Bible nowhere says so, the Jews of the late biblical period considered Jerusalem the center of the "circle of the world"—just as the Greeks thought of Delos as the center. (A spherical surface, of course, has no center.)

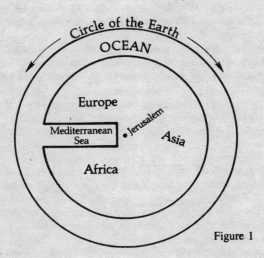

Figure 1

Now let us quote the entire verse:

"It is he that sitteth upon the circle of the earth, and the inhabitants thereof are as grasshoppers; that stretcheth out the heavens as a curtain, and spreadeth them out as a tent to dwell in." (Isaiah 40:22).

The reference to Earth's inhabitants as "grasshoppers" is merely a biblical cliché for smallness and worthlessness. Thus, when the Israelites were wandering in the wilderness, and sent spies into the land of Canaan, those spies returned with disheartening stories of the strength of the inhabitants and of their cities. The spies said:

". . . we were in our own sight as grasshoppers, and so we were in their sight." (Numbers 13:33).

Observe, however, the comparison of the heavens with a curtain, or a tent. A tent, as it is usually pictured, is composed of some structure that is easily set up and dismantled: hides, linen, silk, canvas. The material is spread outward above and then down on all sides until it touches the ground.

A tent is *not* a spherical structure that surrounds a smaller spherical structure. No tent in existence has ever been that. It is, in most schematic form, a semisphere that comes down and touches the ground in a circle. And the ground underneath a tent is *flat*. That is true in every case.

If you want to see the heavens and Earth, in cross section, as pictured in this verse, see Figure 2. Inside the tent of the heavens, upon the flat-Earth base, the grasshoppers that are humanity dwell.

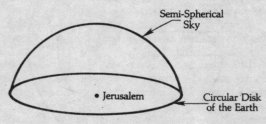

Figure 2

Such a concept is reasonable for people who have not been very far from home; who have not navigated the oceans; who have not observed the changing positions of the stars during travels far north or south, or the behavior of ships as they approach the horizon; who have been too terrified of eclipses to observe closely and dispassionately the shadow of the Earth upon the Moon.

However, we have learned a lot about the Earth and the Universe in the last twenty-five centuries, and we know very well that the picture of the Universe as a tent curtain draped over a flat disk does not match reality. Even Fundamentalists know that much, and the only way they can avoid coming to the conclusion that the Bible is in error is to deny plain English.

And that shows how hard it is to set limits to human folly.

If we accept a semispherical sky resting on a flat-disk Earth, we have to wonder what it rests upon.

The Greek philosophers, culminating in Aristotle (384–322 B.C.), who were the first to accept a spherical Earth, were also the first who did not have to worry about the problem. They realized that gravity was a force pointing to the center of the spherical Earth, so they could imagine the Earth to be suspended in the center of the larger sphere of the Universe as a whole.

To those who came before Aristotle, or who had never heard of Aristotle, or who dismissed Aristotle, "down" was a cosmic direction independent of Earth. As a matter of fact, this is so tempting a view that, in every generation, youngsters have to be cajoled out of it. Where is the youngster in school who, on first encountering the notion of a spherical Earth, doesn't wonder why the people on the other side, walking around, as they do, upside down, don't simply fall off?

And if you deal with a flat Earth, as the biblical writers did, you have to deal with the question of what keeps the whole shebang from falling.

The inevitable conclusion for those who are not ready to consider the whole thing divinely miraculous is to assume the Earth must rest on something—on pillars, for instance. After all, doesn't the roof of a temple rest on pillars?

But then, you must ask what the pillars rest on. The Hindus had the pillars resting on giant elephants, who in turn stood upon a supergiant turtle, which in turn swam across the surface of an infinite sea.

In the end, we're stuck with either the divine or the infinite.

Carl Sagan tells of a woman who had a solution simpler than that of the Hindus. She believed the flat Earth rested on the back of a turtle. She was questioned . . .

"And what does the turtle rest on?"

"On another turtle," said the woman, haughtily.

"And what does that other turtle—"

The woman interrupted, "I know what you're getting at, sir, but it's no use. It's turtles *all the way down*."

But does the Bible take up the matter of what the Earth rests on? —Yes, but only very casually.

The trouble is, you see, that the Bible doesn't bother going into detail in matters that everyone may be assumed to know. The Bible, for instance, doesn't come out and describe Adam when he was first formed. It doesn't say specifically that Adam was created with two legs, two arms, a head, no tail, two eyes, two ears, one mouth, and so on. It takes all this for granted.

In the same way, it doesn't bother saying right out "And the Earth is flat" because the biblical writers never heard anyone saying anything else. However, you can see the flatness in their calm descriptions of Earth as a circle and of the sky as a tent.

In the same way, without saying specifically that the flat Earth rested on something, when everyone *knew* it did, that something is referred to in a very casual way.

For instance, in the thirty-eighth chapter of Job, God is answering Job's complaints of the injustice and evil of the world, not by explaining what it's all about, but by pointing out human ignorance and therefore denying human beings even the right to question (a cavalier and autocratic evasion of Job's point, but never mind). He says:

"Where wast thou when I laid the foundations of the earth? declare, if thou hast understanding. Who hath laid the measures thereof, if thou knowest? or who hath stretched the line upon it? Whereupon are the foundations thereof fastened? or who laid the corner stone thereof." (Job 38:4–6).

What are these "foundations"? It's hard to say because the Bible doesn't describe them specifically.

We might say that the "foundations" refer to the lower layers of the Earth, to the mantle and the liquid iron core. However, the biblical writers never heard of such things, any more than they ever heard of bacteria—so that they had to use objects as large as grasshoppers to represent insignificance. The Bible *never* refers to the regions under the Earth's surface as composed of rock and metal, as we shall see.

We could say that the Bible was written in a kind of double-talk; in verses that meant one thing to the unsophisticated contemporaries of the biblical writers, but that meant something else to the more knowledgeable readers of the twentieth century, and that will turn out to mean something else still to the still more knowledgeable readers of the thirty-fifth century.

If we say that, however, then the entire Fundamentalist thesis falls to the ground, for everything the Bible says can then be interpreted to be adjusted to a fifteen-billion-year-old Universe and the course of biological evolution, and this the Fundamentalists would flatly reject.

Hence, to argue the Fundamentalist case, we must assume the King James Bible to be written in English, so that the "foundations" of the Earth are the objects on which the flat Earth rests.

Elsewhere in the Book of Job, Job says, in describing the power of God:

"The pillars of heaven tremble and are astonished at his reproof." (Job 26:11).

It would seem these pillars are the "foundations" of the Earth. Perhaps they are placed under the rim of the Earth where the sky comes down to meet it, as in Figure 3. These structures are then both the pillars of heaven and the foundations of the Earth.

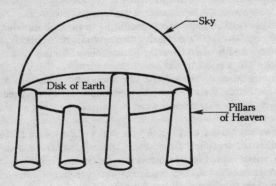

Figure 3

What do the pillars in turn rest on? Elephants? Turtles? Or is it pillars "all the way down"? Or do they rest on the backs of angels who eternally fly through space? The Bible doesn't say.

And what is the sky that covers the flat Earth like a tent?

In the Bible's creation tale, the Earth begins as a formless waste of water. On the first day, God created light and somehow, without the presence of the Sun, caused it to be intermittent, so that there existed the succession of day and night.

Then, on the second day, he placed the tent over the formless waste of waters:

"And God said, Let there be a firmament in the midst of the waters, and let it divide the waters from the waters." (Genesis 1:6).

The first syllable of the word "firmament" is "firm" and that is what the biblical writers had in mind. The word is a translation of the Greek *stereoma*, which means "a hard object" and which is, in turn, a translation of the Hebrew *rakia*, meaning "a thin, metal plate."

The sky, in other words, is very much like the semispherical metal lid placed over the flat serving dish in our fancier restaurants.

The Sun, Moon, and stars are described as having been created on the fourth day. The stars were viewed as sparks of light pasted on the firmament, while the Sun and Moon are circles of light that move from east to west across the firmament, or perhaps just below it.

This view is to be found most specifically in Revelation, which was written about A.D. 100 and which contains a series of apocalyptic visions of the end of the Universe. At one point it refers to a "great earthquake" as a result of which:

". . . the stars of heaven fell unto the earth, even as a fig tree casteth her untimely figs, when she is shaken of a mighty wind. And the heaven departed as a scroll when it is rolled together . . ." (Revelation 6:13–14).

In other words, the stars (those little dots of light) were shaken off the thin metal structure of the sky by the earthquake, and the thin metal sky itself rolled up like the scroll of a book.

The firmament is said "to divide the waters from the waters." Apparently there is water upon the flat base of the world structure, the Earth itself, and there is also a supply of water *above* the firmament. Presumably, it is this upper supply that is responsible for the rain. (How else account for water falling from the sky?)

Apparently, there are openings of some sort that permit the rain to pass through and fall, and when a particularly heavy rain is desired, the openings are made wider. Thus, in the case of the Flood:

". . . the windows of heaven were opened." (Genesis 7:11).

By New Testament times, Jewish scholars had heard of the Greek multiplicity of spheres about the Earth, one for each of the seven planets and then an outermost one for the stars. They began to feel that a single firmament might not be enough.

Thus, St. Paul, in the first century A.D., assumes a plurality of heavens. He says, for instance:

"I knew a man in Christ above fourteen years ago . . . such an one caught up to the third heaven." (2 Corinthians 12:2).

What lies under the flat disk of the Earth? Certainly not a mantle and a liquid iron core of the type geologists speak of today; at least not according to the Bible. Under the flat Earth, there is, instead, the abode of the dead.

The first mention of this is in connection with Korah, Dathan, and Abiram, who rebelled against the leadership of Moses in the time of the wandering in the wilderness:

"And it came to pass . . . that the ground clave asunder that was under them: And the earth opened her mouth, and swallowed them up, and their houses, and all the men that appertained unto Korah, and all their goods. They, and all that appertained to them, went down alive into the pit, and the earth closed upon them: and they perished . . ." (Numbers 16:31–33).

The pit, or "Sheol," was viewed in Old Testament times rather like the Greek Hades, as a place of dimness, weakness, and forgetfulness.

In later times, however, perhaps under the influence of the tales of ingenious torments in Tartarus, where the Greeks imagined the shades of archsinners to be confined, Sheol became hell. Thus, in the famous parable of the rich man and Lazarus, we see the division between sinners who descend into torment and good people who rise into bliss:

"And it came to pass, that the beggar died, and was carried by the angels into Abraham's bosom: the rich man also died, and was buried; And in hell he lift up his eyes, being in torments, and seeth Abraham afar off, and Lazarus in his bosom. And he cried and said, Father Abraham, have mercy on me, and send Lazarus, that he may dip the tip of his finger in water, and cool my tongue; for I am tormented in this flame." (Luke 16:22–24).

The Bible doesn't describe the shape of the pit, but it would be interesting if it occupied the other semisphere of the sky, as in Figure 4.

It may be that the whole spherical structure floats on the infinite waste of waters out of which heaven and earth were created, and which represents primeval chaos, as indicated in Figure 4. In that case, perhaps we don't need the pillars of heaven.

Thus, contributing to the waters of the Flood, were not only the windows of heaven opened wide but, at that time also:

". . . were all the fountains of the great deep broken up . . ." (Genesis 7:11).

In other words, the waters of chaos welled upward and nearly overwhelmed all of creation.

Naturally, if the picture of the Universe is indeed according to the literal words of the Bible, there is no chance of a heliocentric system. The Earth cannot be viewed as moving at all (unless it is viewed as floating aimlessly on the "great deep") and certainly it cannot be viewed as revolving about the Sun, which is a small circle of light upon the solid firmament enclosing Earth's flat disk.

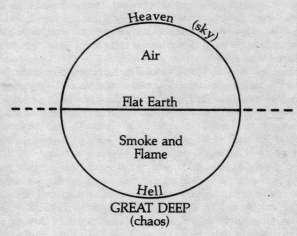

Figure 4

Let me emphasize, however, that I do not take this picture seriously. I do not feel compelled by the Bible to accept this view of the structure of Earth and sky.

Almost all the references to the structure of the Universe in the Bible are in poetic passages of Job, of Psalms, of Isaiah, of Revelation, and so on. It may all be viewed as poetic imagery, as metaphor, as allegory. And the creation tales at the beginning of Genesis must also be looked upon as imagery, metaphor, and allegory.

If this is so, then there is nothing that compels us to see the Bible as in the least contradictory to modern science.

There are many sincerely religious Jews and Christians who view the Bible in exactly this light, who consider the Bible to be a guide to theology and morality, to be a great work of poetry—but *not* to be a textbook of astronomy, geology, or biology. They have no trouble in accepting both the Bible and modern science, and giving each its place, so that they:

". . . Render therefore unto Caesar the things which be Caesar's, and unto God the things which be God's.'' (Luke 20:25).

It is the Fundamentalists, the Literalists, the Creationists with whom I quarrel.

If the Fundamentalists insist on foisting upon us a literal reading of the Genesis creation tales; if they try to force us to accept an Earth and Universe only a few thousand years old, and to deny us evolution, then I insist that they accept as literal every other passage in the Bible—and that means a flat Earth and a thin, metal sky.

And if they don't like that, what's that to me?

XVII

The Armies of the Night

I attended a New York gathering of Mensa last weekend, for I am International Vice-President, and it has become a tradition that I speak at the New York meetings.

Mensa, as you may know, is an organization of high-IQ people and I have encountered many very bright and very lovable people there so that it is an absolute pleasure to attend the meetings.

However, I suspect that it is precisely as easy for a person with a high IQ to be foolish as it is for anyone else.

Thus, a number of Mensans seem to be very impressed with astrology and other forms of occultism, and, on the evening on which I gave my talk, I was preceded by an astrologer who delivered some fifteen minutes of meaningless pap, to my considerable *ennui*.

Moreover, that was not my only encounter with astrology that day.

Mensans have the habit of challenging each other to all sorts of mental combat and I am a natural target for that, although I do my best to avoid it, and to do little more than fend off the rapiers when combat is unavoidable. Or, at least, I try.

On this occasion, a young woman, quite attractive, approached me (knowing who I was, of course) and said, quite aggressively, "Where do you stand on astrology?"

She could scarcely have read much of my writing without knowing the answer to the question, so I gathered she wanted a fight. I didn't, so I contented myself with a minimal statement of my position and said, "I am not impressed with it."

She must have expected that, for she said at once, "Have you ever studied astrology?"

She felt safe in asking that, I suppose, for she undoubtedly knew that a hard-working science writer such as myself is constantly breaking his neck trying to keep up with legitimate science, and that I could scarcely devote much time to a painstaking investigation of each of the many fringe follies that infest the public.

I was tempted to say I had, of course, for I knew enough astronomy to know that astrological assumptions are ridiculous, and I have read enough of the writings of scientists who *have* studied astrology to know that no credence need be given any part of it.

If, however, I had said I was a student of astrology, she would have asked if I had read some nonsensical book by jackass number one, and some idiotic tome by crackpot number two, and she would have nailed me as not only someone who hadn't studied astrology, but who had lied about it.

So I said, with an amiable smile, "No."

She said, promptly, "If you studied it, you might find that you would be impressed with it."

Still responding minimally, I said, "I don't think so."

That was what she wanted. Triumphantly, she said, "That means you are a narrow-minded bigot, afraid to shake your own prejudices by investigation."

I should have simply shrugged, smiled, and walked away, but I found myself driven to a retort. I said, "Being human, miss, I suppose I do have a bit of bigotry about me, so I carefully expend it on astrology in order that I won't be tempted to use it on anything with any shadow of intellectual decency about it." —And she stamped off angrily.

The problem, you see, was not that I had failed to investigate astrology; it was that she had failed to investigate astronomy, so that she didn't know how empty of content astrology was.

It is precisely because it is fashionable for Americans to know no science, even though they may be well educated otherwise, that they so easily fall prey to nonsense.

They thus become part of the armies of the night, the purveyors of nitwittery, the retailers of intellectual junk food, the feeders on mental cardboard, for their ignorance keeps them from distinguishing nectar from sewage.

In a way, though, my astrological adversary left prematurely. She had weapons left in her armory that might easily have lured me into further argumentation of an entirely unprofitable sort.

She might have pointed out that there were great early astronomers who believed in astrology. The great science fiction editor John Campbell once used that argument on me, for instance.

Consider Johannes Kepler! He was an astronomer of the first rank and it was he who first worked out the proper design of the Solar System. —And yet *he* cast horoscopes.

In those days, however, as in these, astrologers earned more money than astronomers did, and Kepler had to make a living. I doubt that he believed the horoscopes he concocted, and, even if he did, that means nothing.

When Campbell used the argument on me, what I said in reply was:

"Hipparchus of Nicaea and Tycho Brahe of Denmark, two of the greatest astronomers of all time, believed the Sun revolved about a stationary Earth. With all due respect to those two authentically great minds, I don't accept their authority on this point."

The young lady might also have said that the Moon certainly affects us by way of the tides and yet for centuries most astronomers scouted that notion. One of their arguments was that every other high tide took place when the Moon wasn't even in the sky.

True enough. And had I lived in the time of Galileo, I might have ignored the influence of the Moon, too, as he did; and I would have been wrong, as he was.

Still, the connection between the Moon and the tides was not an astrological tenet; the existence of that connection was proven by astronomers and not by astrologers; and once the connection was proved it did not lend one atom's worth of credibility to astrology.

The question is not whether the Moon affects the tides, but whether the Moon (or any other heavenly body) affects *us* in such a way as to persuade us that the minutiae of our behavior ought to be guided by the changes in configuration of those heavenly bodies.

We know, you and I, what astrology is. If you have any doubts, read any astrology column in any newspaper and you'll find out. If you were born on a certain day, astrologers say, you should today be careful in your investments, or watch out for quarrels with loved ones, or have no fear of taking risks, and so on, and so on.

Why? What is the connection?

Have you ever heard an astrologer explain exactly why a particular birth date should influence your behavior in some particular way? He might explain that when Neptune is in conjunction with Saturn, financial affairs (let us say) become unstable, but does he ever explain *why* that should be so? Or how he found out?

Have you ever heard of two astrologers arguing seriously as to the effect of an unusual heavenly combination on individuals, with either advancing evidence for his own point of view? Have you ever heard of an astrologer making a new astrological finding or advancing astrological understanding in this respect or that?

Astrology consists of nothing but flat statements. The closest one comes to anything more than that is when someone maintains that the number of (let us say) athletes, born under the ascendency of Mars (or whatever), is higher than is to be expected of random distribution. Generally, even that sort of dubious "discovery" fades on closer examination.

Let's take another example. Some years ago, a book came out entitled *The Jupiter Effect*. It advanced a complicated thesis involving tidal effects on the Sun. Such tidal effects do exist and Jupiter is the prime agent, though other planets (notably Earth itself) also contribute.

Arguments were presented to support the view that these tidal effects influenced solar activity such as sunspots and flares. That, in turn, would influence the solar wind, which, in its turn, would influence the Earth and might, to a minor extent, affect the delicate balance of the Earth's plate-tectonic changes.

As it happened, the planets would be clustered more closely than usual in the sky in March 1982 and their combined tidal effects would be a bit more intense than usual. If the sunspot peak came in 1982, it would be higher than usual in consequence, perhaps, and the effect on Earth would be heightened. *If*, then, the San Andreas fault was on the point of slipping (as most seismologists believe it to be) the effect of the solar wind might supply just that last little straw and bring about an earthquake in 1982.

The authors made no secret that the chain was a long and very shaky one.

The publisher provided me with galleys and asked me for an introduction. I was intrigued by the thesis and wrote the introduction— which was a mistake. I had no idea how many people would read the book and, ignoring the caveats, take it very seriously indeed. I began to be bombarded by fearful letters asking me what would happen in March 1982. At first I'd send back postcards reading, ''Nothing.'' Toward the end the message read, ''Nothing!!!''

As it happened, the sunspot peak came well before 1982, and that spoiled everything. There was clearly no necessary connection between the planetary tidal activity and the sunspot cycle. One of the authors of the book promptly disowned the theory. (And even if he hadn't, all he had claimed was that an earthquake that was going to take place anyway might just possibly happen a little sooner because of the planetary effects—in March, let us say, rather than in October.)

By the time the author disowned the theory, however, it was too late. *The Jupiter Effect* had caught the attention of the armies of the night, and they became enamored of the ''planetary lineup.''

I gathered from the letters I got that they thought the planets would be lined up one behind the other, straight as a ruler. (Actually, they were spread out, even at their closest, across a quarter of the sky.)

They also thought this was an arcane development that happened only every million years or so. Actually such groupings take place about every century and a quarter. In fact, it wasn't many years ago that there was an even closer lineup than the one in March 1982, but on that occasion some of the planets were on one side of the Sun and some on the other.

From a tidal viewpoint it doesn't matter whether planets are all on the same side of the Sun, or distributed on both sides, as long as all are in an approximate straight line, but to the lineup people, only the same side counted, apparently. Having them all on the same side made it seem that the Solar System would tip over, I suppose.

What's more, the planetary-lineup fans weren't content to have an earthquake. The word was that California would slide into the sea.

In fact, even the loss of California wasn't enough for many. The word went out that the world would come to an end, and I presume many people woke up on the day of the lineup all set to meet whatever fate they were counting on when the great THE END appeared in the sky.

I couldn't help but wonder why they bothered to pin the lineup to a single day, by the way. The planets slowly moved across the sky on their separate paths and on one particular day the area within which all were to be found was at a minimum. The day before and the day after, however, the area was very little larger than that minimum, and two days before and two days after, very little larger than that. Whatever the material influence of the lineup, it could not have been very much greater at the moment of minimal area than at any time over a period of several days. I suspect, though, that the lineup addicts had the notion that the whole thing worked through some mystic influence that was exerted only when all the planets slipped behind each other to form an exact straight line (which never happened, of course).

In any case, the day of the lineup came and went and nothing untoward took place.

I knew better than to suspect that a single person would get up and say, "Gee, I was wrong." They're all too busy waiting for the next piece of end-of-the-world chic. Halley's Comet, perhaps.

The illiterates don't even bother to get the vocabulary right.

A theory, when advanced by a competent scientist, is an elaborate and detailed attempt to account for a series of otherwise disconnected and apparently unrelated observations. It is based on numerous observations, close reasoning, and, where appropriate, careful mathematical deduction. To be successful, a theory must be confirmed by other scientists through numerous additional observations and tests and, where this is possible, must offer predictions that can be tested and confirmed. The theory can be, and is, refined and improved as more and better observations are made.

Here are a few examples of successful theories, and the dates upon which each was first advanced:

The atomic theory—1803

The theory of evolution—1859

The quantum theory—1900

The theory of relativity—1905

Each one of these has been endlessly tested and checked since its first advancement and, with necessary improvements and refinements, has passed all challenges.

No reputable scientist doubts that atoms, evolutionary development, quanta, or relativistic motion exist, though further improvements and refinements of details may prove necessary.

What is a theory *not?* It is *not* "a guess."

Many people who know nothing about science will dismiss the theory of evolution because it is "just a theory." No less a brain than Ronald Reagan, in the course of his 1980 campaign, when addressing a group of Fundamentalists, dismissed evolution as "just a theory."

I once denounced one of these "just a theory" fellows in print, stating that he clearly knew nothing about science. The result was that I received a letter from a fourteen-year-old who told me that theories were just "wild guesses" and he knew this because that's what his teachers told him. He then denounced the theory of evolution in unmeasured terms and told me proudly that he prayed in school because no law could prevent him from doing so. And he enclosed a stamped, self-addressed envelope because he wanted to hear from me on the matter.

I felt it only fair to oblige. I dashed off a line asking him to consider seriously whether it might not be possible that his teachers were as ignorant of science as he was. I also suggested that in his next prayer he might implore God to grant him an education, so that he wouldn't stay ignorant all his life.

And that brings up a serious point. How can we keep people from being ignorant, when those who would teach them are so often appallingly ignorant themselves?

Clearly there are flaws in the American educational system, and American schools are particularly weak in science for a number of reasons.

One of the reasons, I suppose, is that good old pioneer tradition that has always held "book learning" in deep suspicion, and felt that good old "horse sense" was all that was really necessary.

That the United States has gotten by and reached world leadership in science and technology has been thanks, in part, to its ingenious tinkerers—the Thomas Edisons and Henry Fords—and, in part, because of the influx of many who had already received European educations or who had absorbed a European respect for learning and saw to it that their children were properly educated.

Adolph Hitler was responsible for literally dozens of top-flight scientists flooding into the United States in the 1930s, and the beneficial effects of their presence and of the pupils they helped develop are still with us and have helped to cushion the inadequacies of American education practices.

This can't continue forever. As our technology grows more complex, it becomes less and less likely that we can depend upon the independent tinkerer. And Hitler's mistake is not likely to be repeated. The Soviets, for instance, go to great efforts not to allow anyone out of their country who might be of use to those they conceive to be their enemies.

Yet over and above the general inadequacies, it would seem that the American school system has deteriorated enormously in the last twenty years. There are horror tales, constantly, of people getting into college without being able to write a coherent sentence. —And it is quite clear to anyone willing to look at the American scene with eyes open that we are rapidly losing our scientific, technological, and industrial leadership.

Why is that? —Here is what I think.

About twenty years ago, the Supreme Court decided that the American Constitution did not allow schools to be segregated on the basis of race, and the courts directed that children be transported out of their neighborhoods to even out black / white ratios. There followed, as we all know, a white flight to suburbs and to private schools, with the result that public schools in most of our large cities are now heavily and increasingly black.

With that, there came a rapid loss of interest in supporting public schools on the part of the white middle class, which supplied most of the financing, and most of the teachers, too.

You must realize that it takes money to teach science well. You need rather elaborate textbooks, elaborately educated teachers, and elaborately equipped laboratories. As the money available decreases, science education suffers disproportionately. Nor does the outlook for the future look anything but bleak. The Reagan administration is steadily cutting support for the public school system and is proposing tuition credits for private schools.

Well, then, you might argue, won't the private schools teach science?

Will they?

The public school system is government-financed. The individual taxpayer cannot easily influence just what his taxes will be spent on, and the school administration, if it has any professional competence, will insist on a well-rounded education. Teachers, as civil servants, are difficult to fire for the crime of thinking, and the Constitution serves to prevent the more egregious abuses against freedom. (This was in the old days, before the public school system was virtually dismantled.)

Private schools, on the other hand, are financed by parents' tuition, and most parents, fleeing a public school system they don't like for whatever reason, cannot easily afford the tuition they must pay on top of their school taxes. Naturally, they don't want to add to the expenses needlessly.

Since an elaborate science education means an additional hike to the tuition, parents are liable to see the virtues of fundamentals, the old, traditional "readin', ritin', and 'rifmetik." That's a fourth-grade education, as it happens, but allowing a *few* frills such as the pledge of allegiance and school prayers, that ought to be enough.

The private schools have to be responsive to the parents and their pocketbooks, so we can look to them for a *safe* education, something that will qualify people for the job of junior executive and develop their ability to handle three martinis at lunch. —But a *good* education? I wonder.

I do not, however, want to divide the world into good guys and bad guys in a simplistic sort of way. Many a nonscientist is intelligent and rational. And, on the other hand, there are scientists, even great ones, who have turned off into the bogs and morasses, both in the past and the present.

It's not really surprising that this is so. The scientific method is an austere and Spartan exercise for the brain. It represents a slow advance at best, gives rise to the Eureka phenomenon both rarely and only for the few—and even for those few, not often. Why shouldn't scientists be tempted to turn away, to find some other route to truth?

I was once a subscriber to a science magazine for high school students, and there came a time when I grew uneasy about it. It seemed to me that the editor was allowing himself to display marked sympathy for Velikovskianism and astrology. Once, when a number of astronomers signed a statement denouncing astrology, the magazine objected, and wondered if the astronomers had really investigated astrology.

I was moved to write a strong denunciation of that silly remark.

The editor responded with a long letter, in which he tried to explain that reason and the scientific method were not necessarily the only routes to truth and that I should be more tolerant of competing methods.

That irritated me. I sent him a rather brief letter that went (as nearly as I can remember) something like this: "I have your letter in which you explain that reason is not the only route to truth. Your explanation, however, consists entirely of an attempt at reasoning the point. Don't tell me; *show* me! Convince me by dreaming at me, or intuiting. Or else write me a symphony, paint me a painting, or meditate me a meditation. Do something—*anything*—that will place me on your side and that isn't a matter of reasoning!"

I never heard from him again.

Here's something else. Some months back, *Science Digest* was planning to publish an article about various present-day top-rank scientists, including Nobel laureates who have developed odd and mystical notions about the human mind, who are trying to penetrate the secrets of nature by meditation, who are strongly influenced by oriental philosophies, and so on. *Science Digest* sent me the manuscript and asked for my comments.

I wrote a letter in response, which was included in a box (under the heading of "Science Follies") and which accompanied the article that was published in the July 1982 issue of the magazine. Here is the letter, word for word:

"Throughout history, many great scientists have worked on some farfetched ideas. Johannes Kepler was a professional astrologer. Isaac Newton tried to change baser metals into silver and gold. And John Napier, who invented logarithms, devised a monumentally foolish interpretation of the Book of Revelation.

"This list goes on. William Herschel, the discoverer of Uranus, thought the sun was dark, cool, and habitable under its flaming atmosphere. The American astronomer Percival Lowell insisted he saw canals on Mars. Robert Hare, a very practical American chemist, invented a device by which he could communicate with the dead. William Weber, a German physicist, and Alfred Wallace, co-deviser of the modern theory of evolution, were ardent spiritualists. And the English physicist Sir Oliver Lodge was a dedicated supporter of psychic research.

"Knowing this track record, I would be enormously astonished if, in the year 1982, there were suddenly no great scientists who fell in love with speculative notions that seem, to lesser minds such as mine, to be irrational.

"Unfortunately, most of these speculative theories cannot be tested in any reasonable way, cannot be used to make predictions, and are not presented with compelling arguments that could convince other scientists. In fact, among all these devoted imagineers, no two entirely agree. They doubt one another's rationality.

"It may be, of course, that out of all this apparent nonsense, some nuggets of useful genius will tumble out. That such things have happened before is enough to justify it all. Nevertheless, I suspect that these nuggets will be few and far between. Most of the speculations that seem to be nonsense—even when great scientists are the source—will, in the end, turn out to be nonsense."

So there you are. I stand four-square for reason, and object to what seems to me to be irrationality, whatever the source.

If you are on my side in this, I must warn you that the army of the night has the advantage of overwhelming numbers, and, by its very nature, is immune to reason, so that it is entirely unlikely that you and I can win out.

We will always remain a tiny and probably hopeless minority, but let us never tire of presenting our view, and of fighting the good fight for the right.

NEW AVON ● DISCUS TITLES

GREAT AMERICAN POLITICAL THINKERS,   83915-6/$4.95
VOL. I & II Bernard E. Brown, Ed.   83823-7/$4.95
These volumes present a comprehensive survey of American
political thought from the early 17th century to the present.
Thomas Jefferson, Alexander Hamilton, Abraham Lincoln, H.D.
Thoreau, Oliver Wendell Holmes and Hannah Arendt are among
those whose writings are anthologized here, illuminating the
key issues of their time.

COWARD THE PLAYWRIGHT John Lahr   64683-8/$3.95
Noel Coward—actor, performer and playwright,—is best known
for his witty and sophisticated comedies of the British
"leisure class." This intriguing, impeccably researched
biography illuminates the work and philosophy of one of the
greatest playwrights and most enigmatic personalities of our
time. 8 pages of photographs.

LAST WALTZ IN VIENNA George Clare   64907-5/$3.95
This extraordinary, evocative account documents the rise
and fall of a Jewish family in Austria. "A brilliant
interweaving of the personal and the public...We never
lose sight of the historical context in which the family's
destiny unfolded." *The New York Times*

RETHINKING LIBERALISM   84848-1/$4.95
Walter Truett Anderson, Ed.
This provocative collection of essays confronts the current
crisis of liberal ideology and addresses the vital issues now
reshaping liberal politics in order to meet the challenging
political realities of the 1980's: health care, education,
employment, welfare, crime, foreign policy, environmentalism,
biopolitics, energy and global interdependence.

PEOPLE OF THE LAKE   65938-7/$3.95
Richard E. Leaky and Roger Lewin
A fascinating look at man's origins—and their implications
for human destiny—by Richard Leaky, the great paleoan-
thropologist. "Profound in its implications."
Washington Post Book World

AVON Paperbacks

NEW AVON ⬤ DISCUS TITLES

FANTASISTS ON FANTASY
Robert H. Boyer and Kenneth J. Zahorski   86553-X/$3.95
This volume offers a glimpse into the world of eighteen
of the world's premier fantasy writers, with twenty-two
pieces—ranging from critical essays to personal letters—
in which the real experts in the field explore the theory,
the technique and the aesthetics of fantasy literature.

COUNTING THE EONS Isaac Asimov   67090-9/$3.95
From "the greatest explainer of the age" (Carl Sagan)
comes an "entertaining, witty and informative"
(*Library Journal*) collection of seventeen essays
covering the most fantastic phenomena of the universe
from robots to Einstein, black holes to anti-matter.

KRISHNAMURTI: The Years of Fulfillment
Mary Lutyens   68007-6/$4.95
This "extraordinary document" (*San Francisco Chronicle*),
is the second volume in the illuminating biography—
begun in KRISHNAMURTI: The Awakening Years—of one of
the most radiant spiritual personalities of all time.

FATAL FLOWERS: On Sin, Sex and Suicide in the Deep South
Rosemary Daniell   65946-8/$3.95
Here is a shockingly intimate, unsparing portrait
of women destroyed by a society drenched in sexual
obsession and Bible-belt repression. "One of those fine
female writers unique in her region (Eudora Welty,
Flannery O'Connor, Katherine Anne Porter, Alice Walker,
Caroline Gordon)." *Time*

AVON PAPERBACKS